TRANSIENTS, SETTLERS, AND REFUGEES
Asians in Britain

Transients, Settlers, and Refugees

Asians in Britain

VAUGHAN ROBINSON

CLARENDON PRESS · OXFORD

1986

Oxford University Press, Walton Street, Oxford OX2 6DP
Oxford New York Toronto
Delhi Bombay Calcutta Madras Karachi
Kuala Lumpur Singapore Hong Kong Tokyo
Nairobi Dar es Salaam Cape Town
Melbourne Auckland
and associated companies in
Beirut Berlin Ibadan Nicosia

Oxford is a trade mark of Oxford University Press.

Published in the United States
by Oxford University Press, New York

British Library Cataloguing in Publication Data
Robinson, Vaughan
Transients, settlers, and refugees: Asians in Britain.
1. South Asians—England—Blackburn (Lancashire)
—Social conditions   2. Blackburn (Lancashire)
—Social conditions
I. Title
305.8'914'0427623   DA690.B63
ISBN 0-19-878009-5

Library of Congress Cataloging in Publication Data
Robinson, Vaughan
Transients, settlers, and refugees.
Bibliography: p.
Includes index.
1. Asians—England—Blackburn (Lancashire)
2. Blackburn (Lancashire)—Foreign population.
3. Asians—Great Britain—Case studies. I. Title
DA690.B63R63   1985   305.8'009427623   85-18858
ISBN 0-19-878009-5

Set by Colset Pte Ltd, Singapore
Printed in Great Britain by
Billings & Sons Ltd, Worcester

To
Mum and Dad

PREFACE

The roots of this book go back as far as 1975, when as a second-year undergraduate at St Catherine's College I produced a dissertation on the simulation of 'ghetto' growth in Blackburn. Like many young geographers at the time I was in awe of quantification, the computer, and the scientific route to explanation. These themes were given strong emphasis in my application for an SSRC postgraduate award based on an extension and development of the dissertation and its methodology. However, even as my application was successful I began to have doubts. My reading of David Ley's work on gang turfs and graffiti, David Smith's utterances about relevance and welfare geography, and David Harvey's iconoclastic treatises all suggested that simulation modelling was no longer at the heart of the 'New Geography'. These feelings were emphasized by a term of being supervised by Jean Gottmann. Not once did he mention either simulation or computers. Instead we talked of anthropology.

The transformation was complete when, in 1978, I took up a studentship at Nuffield College, Oxford. By this time I had all but decided to abandon the simulation element and the scientific method, and instead be rather more pragmatic about my methodology. The subject matter, the Asian population of Blackburn, remained but my approach, methods, and outlook had changed. In particular, under the guidance of Ceri Peach I had begun to explore the sociological literature more fully and had found it both more profound and more relevant than equivalent geographical writings. My horizons were broadened even further when I was allocated to Clyde Mitchell, the eminent anthropologist, for college supervisions at Nuffield. Whilst he admired the work of some geographers and freely admitted the importance of place and space, Clyde was deeply sceptical of geography and its 'methods'. He encouraged me to read more widely in anthropology and sociology and forced me to justify explicitly many of the assumptions which I made and which would have gone unchallenged in a purely geographical audience. Supervisions with Clyde were often dispiriting and unnerving but they were always valuable and they taught me a great deal. Ironically they also taught me the very real strengths which geography has as a subject when allied to those of sociology and anthropology. Other Fellows at Nuffield contributed indirectly to the substance of my doctorate and subsequently to this manuscript, although they rarely realized it. The presence of John Goldthorpe, Chelly Halsey, Jim Sharp, and Clive Payne amongst others combined to create an atmosphere which was stimulating, dynamic, vital, and within which it was difficult not to give of your best. Without my four years at Nuffield (latterly as a Research Fellow) this book would never have materialized.

My last year at Nuffield also saw me undertaking a field visit to the sending areas in India and Pakistan. For the insight and information which I gained from this trip I am indebted to the Unilever Corporation and in particular to the company secretary, James Kier. The precision and efficiency with which Unilever, and its subsidiaries Hindustan Lever and Lever Brothers Pakistan, made all the arrangements for this visit merely served to confirm my admiration for this magnificent company. My particular thanks go to Dr Ganguly, Saleem, Atul, Mahesh, and all the many villagers who invited us into their homes, talked so freely, and offered us their hospitality so generously.

At the opposite end of the migration stream, it would have been impossible to write a book about Asians in Blackburn without the wholehearted support of that local authority and its appointed officers. Clifford Singleton, the Chief Executive, members of the Planning, Housing and Rating Departments, and Len Proos the Senior Community Relations Officer (SCRO) were all knowledgeable, patient, and helpful. My particular thanks go to the Asian interviewers who worked on the Job Creation Project which collected much of the data on which this study is based.

The end product of these various elements was my doctoral thesis which was successfully submitted in 1982 just prior to my departure from Oxford. Colin Clarke, my internal examiner, was constructive in his criticism and invaluable in helping me get this book published.

My move to University College, Swansea caused a research and domestic dislocation which has delayed the publication of this book considerably. This has not been without benefits though. The hiatus allowed me to stand back from earlier drafts and see their obvious weaknesses. The need to produce teaching material for my course on the Geography of Race Relations has given me the opportunity to read widely and think in broader terms. I now see the context within which my own work lies, whereas previously I did not. It is hoped that the delay in publication has not dated this work too much, and wherever possible it has been updated using 1984 data. Inevitably though, the large-scale 1977 Asian Census which forms the empirical backbone of the study could not be repeated.

My move to Swansea has also meant that the production of this book has taken place within two separate departments. I am indebted to my secretary at Nuffield, Sharon, and to the team at Swansea, Glenys and Mary. I also owe a debt to the cartographic unit at Swansea, and in particular to Tim Fearnside.

Finally, there are the constants in the equation which has produced this book: Ceri Peach, who has at various times been my undergraduate tutor, my postgraduate supervisor, and research colleague. Over a period of eight years Ceri guided, encouraged, supported, advised, and helped me in a manner more akin to an academic father than a mere teacher; Jayne, my wife, who has uncomplainingly read and commented on article after article, and chapter after chapter of my work, and who has talked patiently to me about 'encapsulation', 'marginality', and all the other concepts which seemed to be so vitally important to me at 2 o'clock in the morning. I hope this book is some consolation to her for the lost sleep, the non-existent weekends and the even rarer holidays; lastly there are my parents who have devoted their energies and lives to ensuring that my brother and I had a better start in life than they did. Perhaps the dedication of this book to them will repay a little of this sacrifice, love and confidence.

Glais   Vaughan Robinson
July 1984

ACKNOWLEDGEMENTS

I am grateful to Penguin Books Ltd. for permission to reprint sections of Jeremy Seabrook's *City Close Up* first published in 1971.

CONTENTS

LIST OF FIGURES

CHAPTER 1

Introduction

The study of ethnic settlement in British cities cannot yet claim an academic pedigree which stretches back over many decades and which is based on a coherent body of knowledge and literature. Published research has been sporadic in nature and shows little evidence of incremental progress towards a commonly acknowledged goal. Six factors can be put forward to explain this state of affairs.

Firstly, fine scale census data have only recently become available to researchers. Enumeration District-level analysis only became possible after 1961, and sources of street-level information have only been recognized in the 1970s. Consequently, detailed micro-scale analyses are still not commonplace.

Secondly, the data which are accessible to researchers are bedevilled by inconsistencies, omissions, and inaccuracies. A case in point is the numerous attempts which officials make to quantify ethnic status: the census, for example, uses birthplace and a national sample whereas the National Dwelling and Housing Survey allowed respondents to identify their own ethnicity but was only based on a six per cent sample of the population. Given such confusion over exactly what is being measured, it is inevitable that researchers using these sources should have become involved in non-productive disagreements about definitions and interpretations.

Thirdly, the study of urban social patterns in the UK has never been dominated by a pre-eminent school of thought as was the case in the United States with the Chicago School of Park, Burgess, *et al.* Whilst some would argue that British ethnic studies have contained distinct schools such as Michael Banton's 'Bristol School' or John Rex's 'Aston School', none of these has ever been sufficiently attractive or persuasive to dominate thinking. The British literature consequently relies upon a large number of individualistic and weakly developed paradigms, and it has at times become the battlefield for internecine warfare between the proponents of different stances.

Fourthly, the history of race relations in the UK is one of stifled rather than open conflict and tension. Until the spectacular growth of the National Front in the mid-1970s (Taylor, 1982; Robinson, 1985a) and the summer riots of 1980 and 1981, the social pressure simply did not exist to stimulate persistent and serious research into the issue.

Fifthly, and perhaps of lesser importance than the other four reasons, was the absence of learned journals which dealt specifically with ethnic relations as an issue. *Race* was one of the earliest to adopt this mantle but its content was international and its coverage stressed breadth not depth. The political struggles within the Institute of Race Relations signalled its demise as a serious force. *New Community* was launched in 1972 but has only really come of age in

1

the 1980s. At the close of the 1970s two further avenues of publication became available for the dissemination of research findings, *Ethnic and Racial Studies* and *Immigrants and Minorities*. Together, these three journals are likely to strengthen the position of ethnic studies in Britain and increase the awareness of its importance within the academic and policy-formulating communities.

Coupled with these issues, which relate specifically to British circumstances, there is also the more general problem of the disciplinary status of ethnic studies. The topic suffers from being essentially fringe territory in all of those many academic disciplines which claim sovereignty over it. In Britain, whilst there have been a few notable scholars who have pursued the theme as their central interest, much of our current knowledge has been produced by researchers who have come into the field whilst studying broader or related issues.

Within the British literature, then, these six trends or problems have combined to produce work which has tended to be diverse in nature, but above all else, primarily descriptive and empirical. Theory which can stand close scrutiny and claim to be unique to ethnic studies is still a rarity. Because of this, it is easy to be misled by the burgeoning literature and to assume that quantity of material signifies understanding, explanation, and academic maturity. On the contrary, an analysis of the content of this output reveals a sub-discipline which is still in its infancy and which is preoccupied with the earliest stage of intellectual rigour, namely the accumulation of empirical data. The subsequent process of organizing this material through systematic classification has hardly begun, and because of this, the meaningful comparison of differing case-studies or differing outcomes still seems to be beyond the academic horizon.

The inner city riots which occurred throughout Britain during the summers of 1980 and 1981, and the recurrent signs of tension witnessed, for example, by racial harassment and the existence of ethnic vigilante groups, suggest that academic endeavour has remained too long in the descriptive and empirical phase. It suggests, also, that the claim which workers in the field have to direct social relevance is increasingly coming under pressure as academics consistently prove themselves incapable of predicting such social disturbances, explaining their causes, or agreeing on courses of remedial action. In short, the study of ethnic relations in Britain is faced with the challenge of accelerating its academic maturation or facing the same collapse of confidence in its ability which beset other branches of the social sciences in the harsh social and economic climate of the 1970s and 1980s. Whilst this challenge may represent an opportunity both to escape the stigma which is now attached to other parts of the social sciences, and to reassert claims to academic respectability (and, more importantly some would say, to central government funding), it does also offer the spectre of failure. Practitioners are therefore faced with the unenviable task of choosing between slow, incremental progress through selective classification and the search for particularistic explanations which are appropriate only to carefully and closely delineated conditions, or the more direct shift to universalistic criteria and explanations with all the dangers which go with these.

This book represents a modest attempt to blend both the particular and the universal. It aims at one level to propose general explanations about the position of Asian migrants in British society and, in so doing, its subject matter and methodology deliberately remain unrestricted by disciplinary boundaries. The search for theoretical insight consequently ranges over sociological, geographical, anthropological, and to a certain extent historical territory. This conscious policy has been pursued in an effort to turn what has proved, until now, to be the sub-discipline's major weakness into a strength, in that each discipline contributes its own approach and techniques, but strongly directed towards a common focus. However, having pointed out that the text is concerned with explanations and theories which possess some general appeal, these are at all times tempered by a persistent concern with detail and context. At a second level, then, the book is about the particular, and the way in which the case-study material presented here relates to other community studies with their unique contexts and conclusions. In fact, one of the strongest themes running through the study is the need to develop classifications which distinguish between circumstances in contrasting types of towns or between the attitudes and reactions of different groups within Britain's Asian population. In consequence, the book does propose theoretical explanations of some generality but it also considers the way in which the local environment and the characteristics of the migrant populations impinge upon these general explanations to produce local outcomes. Moreover it is argued that the local outcomes which have been produced in Blackburn are not utterly unique to that town, but are characteristic of a certain 'type' of town, and that this category takes in a sizeable number of other settlements of which Blackburn may be thought to be representative.

Furthermore, it is important to note that this group of settlements which shares common characteristics with Blackburn is, and has been, under-researched relative to the multi-racial areas of Britain's largest conurbations. It could be argued that this over-emphasis on the race relations climate in these latter circumstances has been responsible for creating unrepresentative and unfavourable stereotypes in the minds of both policy makers and the general public. Again, the emphasis of this book on a smaller town with an ageing but relatively plentiful housing stock serves to counter-balance the tendency to inflate the problems of Birmingham or Brixton into the problems of all multi-racial settlements.

The Structure of the Book

It is one of the central contentions of this book that a study of the behaviour of Asian immigrants in isolation from its social, economic, and residential context would prove to be only a descriptive and not an explanatory exercise. Consequently the early part of the book is concerned with mapping out and detailing this context. Chapter 2 looks at the urban hierarchy in the UK and the way in which the black population has only settled within certain types of urban area whilst avoiding others. A typology of ethnic settlement is presented which allows the classification of individual cities, such as Blackburn, into groups which share common characteristics. In this way, it is possible to develop a profile of the opportunities which any particular type of settlement offers coloured immigrants and also the constraints which shape the lives of all the residents of such towns. Clearly, this economic and social context has a direct bearing upon the level of potential conflict which exists between immigrants and local residents. Chapter 3 looks at the migration process itself and the reasons why different Asians groups left the Indian sub-continent and East Africa to live and work in the UK. Again this historical and demographic context is likely to be a major determinant of the strategies which different groups adopt towards life in Britain. Chapter 4 looks at less tangible aspects of the context into which Asians migrated, namely the attitudes of the white population and their reaction to the presence of coloured migrants. The way in which these attitudes become embodied in behaviour is illustrated using the example of housing and access to this resource. Finally, Chapter 5 looks at the sending societies, the attitudes which exist within them and the economic, social, and environmental conditions which prevail there. These again have a direct bearing upon the aspirations and expectations of migrants on arrival.

Chapter 6 draws together the arguments presented in previous chapters and consolidates these into an holistic model that provides a simplified representation of British society and the place of coloured immigrants within it. This model identifies certain research hypotheses which are taken up in subsequent chapters.

Chapter 7 thus looks at how the Asian population of Blackburn came to reside in Britain, and how the two main groups (South Asians and East African Asians) are likely to adopt different roles in British society because of their contrasting migration histories.

Chapter 8 is concerned with the relations which exist between the local white population of Blackburn and the Asians who also live in the town. In particular it discusses whether whites do engage in exclusionary closure and whether Asians pursue usurpationary tactics.

Chapter 9 takes up the issue of whether South Asians and East African Asians react in a similar manner to the exclusion with which they are faced or whether they adopt different strategies to achieve their specified goals.

Chapter 10 focuses only upon differences within the Asian population in an effort to develop a continuum of ethnic association which stretches from those groups categorized as traditional (or encapsulated) through to those which are labelled non-traditional.

Chapter 11 contains an interpretation of the evidence and some thoughts on future race relations in Britain and how these might be improved.

CHAPTER 2

A Typology of Black Settlement in Britain

Introduction

Abler, Adams, and Gould (1971) in their book *Spatial Organisation: the Geographer's View of the World* provide a most succinct and perceptive description of the stages through which the discipline of geography has developed over the last 2200 years. In many ways their description has validity for disciplines other than geography and one can certainly argue that the analysis of ethnic settlement has moved along a very similar track in its sixty-year existence in the US and its thirty-year existence in the UK. Clearly though, with the benefits of hindsight, the emergence of survey methodology as a science, and the development of computational hard- and soft-ware, the speed with which the sub-discipline has progressed is of a very different order.

According to Abler, Adams, and Gould the initial phase of geographical enquiry was concerned primarily with one major issue. This was 'where' things were in absolute space, and it entailed the accumulation of data through voyages and expeditions of discovery, and its presentation on accurate maps. Phase two differed significantly in that geographers no longer simply asked 'Where?' but 'What is where?' Researchers had reached a point of marginal returns and the simple accumulation of data began to appear rather less important than how these data could be organized to yield maximum understanding. The mode of enquiry which this phase produced was the regional method in which the researcher was taught to study small and discrete areas in exhaustive detail—relief, drainage, climate, agriculture, economic production, and population. At a later stage, others elected to approach the data matrix from a different angle and, in preference to studying all aspects of a chosen region, they considered a chosen topic throughout all regions. Systematic geography was born, and with it, the attempt to classify areas into coherent groups or types which shared some common characteristics. Clearly these early phases of describing the location of places, the characteristics of places, and how different places could be classified and categorized, formed essential prerequisites to the subsequent phases concerned with model building, relative location, and the explanation either of structures or of individual behaviour.

This digression into a discussion of the development of geography as a discipline is important because it provides a scale against which one can compare the relative academic maturity of the study of ethnic settlement. Abler, Adams, and Gould's framework can also be used as a pointer to the directions which the specialism might take next.

Like geography, the study of ethnic settlement began with a period of fact finding in which the interests of practitioners centred largely upon enumerating the ethnic groups which were present at the national level and describing their distribution in absolute space. In the main these were static studies in the sense that they formed a snapshot of a particular ethnic group or country at one point in time. There is a wide variety of articles which could be used as examples of this phase of research but a number stand out as being particularly important. Hartshorne's (1938) attempt to describe the distribution of various ethnic groups throughout the USA is a classic of its type. Calef and Nelson's (1956) study of the negro population in the US is another, whilst in the UK, Peach's (1966) description of the distribution of West Indians is in the same vein. Peach's (1968) later study of patterns of West Indian settlement within cities provides an additional British example, although here the concern is with finer and more precise areal units. Research such as this, then, formed the first phase in the development of the study of ethnic settlement. It was concerned with answering the simple question 'Where?' about ethnic groups. Where do negroes live within the US? Where do West Indians live within British cities?

Soon, however, the social geographer became concerned not only with 'Where?' but also with 'What is where?'. In geography, as we have seen, this phase produced the regional method: in the study of ethnic settlement it produced the local community study. Community studies either looked at one specific location or sometimes even at one ethnic group in one location. They tried to answer the questions 'Who are these people?' and 'How do they live?'. A representative example of this particular approach is provided by Davison's (1966) study of the coloured population in London. The chapter headings and their contents provide an excellent description of the issues with which the community study was concerned.

The immigrant population of London: who are these people?; where did they come from?; how many came?

Housing conditions: where and how do immigrants live?

Employment: what jobs do immigrants do?; where do they work?; how much are they paid?

Unemployment: how many immigrants are unemployed?

Household budgets: on what do immigrants spend their money?

Migration intentions: are the immigrants going to stay in Britain or return 'home'?

Other researchers pursued broadly similar methodology either at the national level (Rose, 1969) or at the very local level (Jones and Davenport, 1972).

Immediately after this phase of the sub-discipline's development, research methodology progressed rapidly to the point where an incremental accumulation of knowledge through steady progress along commonly recognized lines of advance broke down. Where *geographers* had shifted from collecting information about places to analysing the same data in a systematic fashion, the study of ethnic settlement took a quantum leap from community studies into behaviouralism, humanism, structuralism, modelling, and the quantification of residential location in relative space. Each of these approaches has proved fruitful in its own way but each has its own body of theory, techniques, and views on which issues are worthy of investigation.

Consequently, this fission of research approaches has left the sub-discipline without an integrative focus and has denied it the vital classificatory phase which makes sense of the empirical and descriptive material so laboriously collected in the earlier stages. It seems remarkable that, some twenty years after large-scale coloured migration to the UK, there have been so few attempts to draw together the welter of often conflicting community studies into a classificatory typology. Practitioners, for example, have argued extensively about the value of the housing class concept—some have undertaken local research and argued that it has no validity beyond the Sparkbrook case-study first presented by Rex and Moore. Others suggest that it has universal applicability. It seems a trite point, but what has been almost totally overlooked is that Sparkbrook is a different kind of social and residential environment from Newcastle, London, or Blackburn. There may well be a group of cities or boroughs which share common characteristics with Sparkbrook and where the housing class concept is therefore a useful analytical tool. Conversely, there is a variety of contrasting groups of towns which differ in some important way from Sparkbrook and where, as a result, the concept seems to lose its *raison d'être*. It does seem trite to point this out, but researchers into ethnic settlement have either not realized this fact or have ignored it. As a direct consequence, the sub-discipline has lost its analytical potency and much of its immediate relevance and it has become obsessed with arguing the validity of individual conclusions derived from specific case-studies rather than incorporating such conclusions into wider schemata. The remainder of this chapter seeks to provide such a schema in that it focuses on constructing a typology of ethnic settlements taking into account prevailing local conditions. It is hoped that this framework may at last bring together the community studies and perhaps elevate their value from being descriptive to being explanatory. If it also encourages a more constructive attitude in the sub-discipline then this would also be a worthwhile achievement.

The Urban Hierarchy and Ethnic Settlement

The scale and form of the ethnic hierarchy

Jones (1978), in a very significant article on the distribution and diffusion of Britain's coloured population, was one of the first to attempt to draw together data on the pattern of ethnic settlement at a national level. His research represents a tentative step towards classification and explanation and provides the foundations on which this section of the book rests. Using 1971 census data, Jones argued that the coloured population had developed its own urban hierarchy and that this did not follow that of the total population. The two did share a common characteristic in that Greater London represented a primate city and that, in both cases, Birmingham was the second largest centre but beyond this, the two rank orders diverged. Wolverhampton, for example, which ranked only fifteenth in terms of its total population was the third largest centre of black settlement whilst High Wycombe which ranked one hundred and twentieth on total population was twenty-fourth on coloured population. Other towns or cities exhibited the opposite tendency: Sheffield, Bristol, and Manchester all had a smaller black population than their total population might have suggested.

Jones developed the analysis by segmenting the hierarchy into different 'orders' of ethnic settlement such that Greater London and Birmingham were 'first order' centres with coloured populations in excess of 90,000. Leicester, Bradford, Wolverhampton, Manchester and Coventry constituted the 'second order' with coloured populations of between 20,000 and 30,000. Towns and cities such as Leeds, Nottingham, Luton, Bolton, and West Bromwich formed a 'third order' of ethnic settlements with between 8000 and 16,000 coloured residents, and the remaining centres such as Reading, Blackburn, and Bedford made up a 'fourth order', each with a coloured population of about 6000.

Jones then attempted to create what he termed an 'explanatory framework' for these patterns, and in doing so, he elected to concentrate upon the issue of employment opportunities. He reiterated Peach's (1967) argument about coloured immigrants acting as a replacement labour force in marginal employment, and suggested that, on the basis of the 1971 census data such an explanation appeared still to be of relevance. In Jones' words (1978:521):

coloured labour was employed in circumstances where, for whatever reason, white labour was scarce or unobtainable. Thus, in the case of the textile industries, in which employment and output have been contracting overall, low wages and shift working have made it difficult to recruit white labour. In the expanding regions of south-east England and the east and west Midlands, manufacturing industries . . . experienced chronic labour shortages and public services found great difficulty in filling low-paid posts . . . The major features of the distribution of the coloured population in 1971 were therefore controlled by differential opportunities for employment; the presence of col-oured immigrant labour was related directly to the degree of difficulty experienced in recruiting white labour . . .

Jones did go on to discuss other aspects of coloured settlement, most notably demographic characteristics, but his major contribution from our point of view was his realization that ethnic settlements formed a separate and coherent urban system and that this might be generated by a different combination of causal factors. However, since Jones relied upon (unpublished) 1971 data, it is important to update his empirical findings by selective use of the 1981 census, before continuing further analysis.

With the abortion of the proposed question on ethnicity in the 1981 census, published data from this source consequently still rely on birthplace, and can therefore only relate to immigrants to this country. The so-called black British are therefore categorized, not with their parents, but with white Britons. In addition, the period between the 1971 census and its successor witnessed the Redcliffe–Maud report and the resultant reorganization of local authority boundaries. These two changes mean that the 1981 data are both less represen-tative of the coloured population as a whole, and not comparable on spatial scale with the 1971 data unless they are replotted into grid squares. It goes without saying therefore that temporal comparison is fraught with difficulties and cannot be regarded as at all precise. Having said this, one can draw some comparisons of value at the level with which we are concerned.

At a broad scale, the 1981 urban hierarchy appears very similar to the 1971 hierarchy which Jones outlined. Greater London still dominates the picture with a New Commonwealth and Pakistani population (NCWP) of 630,859 (43 per cent of the total), but local government reorganization now sees the consolidation of the smaller centres into the new Metropolitan Counties. West Midlands contains an NCWP population of 193,855 (13 per cent), West Yorkshire contains 71,568 (5 per cent), Greater Manchester contains 64,008 (4 per cent), South Yorkshire 15,869 (1 per cent), and Merseyside 10,800 (0.7 per cent.) In total then these six major conurbations housed over 66 per cent of the NCWP born population. Table 2.1 breaks down these conurbations into their constituent sub-areas to provide data roughly comparable with that published by Jones. It is noteworthy that a rank correlation of the 1971 and 1981 data produces a coefficient of plus 0.94. Clearly the overall pattern of the hierarchy has changed very little. However there do appear to be some important changes, with many of the textile towns (Blackburn, Bolton, Oldham, and Bradford) moving up the hierarchy, and many of the Midlands centres moving down (Wolverhampton, Coventry, Nottingham, Walsall, and Derby). This supports Jones's notion that the major growth centres during the 1970s would be those where the ethnic population was undergoing family reunion as wives and children entered the country to join their families. According to the census the four textile towns mentioned above had an average of 1876 coloured males per 1000 coloured females in 1971 whereas the ratio for the five Midlands towns was 1229:1000. Clearly the latter were considerably more mature settlements in demographic terms at the start of the decade and could therefore expect a slower rate of in-migration.

This somewhat differential growth in the immigrant populations of the larger ethnic settlements may not have upset the rank order too greatly but it has caused a re-crystallization of the hierarchy into new segments. London and Birmingham are clearly 'first order' settlements with NCWP populations in excess of 85,000, Leicester and Bradford form a 'second order' with 30,000–40,000 immigrants, whilst the cities between Wolverhampton and Leeds in the ranking create a 'third order' with populations of between 17,000 and 20,000. Beneath these, a fourth and fifth order can be discerned with 9000–14,000 NCWP residents (Luton to Blackburn in the ranking) and 6000–7000 immigrants respectively.

Whilst it is undoubtedly of value to discover that the coloured population has developed its own urban hierarchy and that this does not slavishly follow that of the total population, this in itself is insufficient. Jones discussed some of the characteristics of this hierarchy and in particular we have described how he discovered the number of urban areas in the ethnic hierarchy and the way in which the total coloured population was distributed between these. What Jones did not do, though, was to discuss the characteristics of the urban areas themselves and therefore what particular feature or features might be important in determining which centres attracted a coloured population and which did not.

Britain's urban hierarchy

As yet there has been no significant research on the causes of the hierarchy in Britain's ethnic settlement, so one is forced to rely upon the more general literature on urban typologies. This literature is itself not extensive but it does contain a number of comprehensive and thought-provoking contributions which are sufficient to provide us with a sound framework within which a typology of ethnic settlements can be constructed.

Moser and Scott (1961) are rightly considered to be the instigators of this line of research in the UK. Their objectives were 'first to assemble and collate material, pointing out the similarities and contrasts, and secondly to classify towns on the basis of their social, economic and demographic characteristics' (Moser and Scott, 1961:2). They chose to undertake this analysis only on towns which had populations greater than 50,000 in 1951, and therefore narrowed their focus to 157 settlements in England and Wales. Their data set consisted of sixty variables, largely drawn from the 1951 national census although this was supplemented as a source by Board of Trade and Ministry of Housing statistics. The data contained inputs which related to population size, structure, and dynamics, housing and household characteristics, social and economic character, political allegiances, and the provision of health and educational services. These factors, or the variables which quantified them, were then subjected to principal components analysis. In the case of Moser and Scott, the technique collapsed sixty variables into only four major components which between them accounted for some 60 per cent of the total variance. Moser and Scott then derived component scores for each of their towns on each of the four components. They then grouped together those towns which had produced similar scores on the first two of the four components, and lastly, assigned those towns which appeared marginal on this criterion by reference to their scores on the other two components. A subsidiary requirement was also built into the assignment procedure to produce groups which had less than ten towns within them. The results of this analysis are summarized in table 2.2, but in brief, Moser and Scott succeeded in dividing up the main towns and cities of England and Wales into three major types which between them contained thirteen categories. Although the detail of Moser and Scott's methodology has been criticized, most notably by Andrews (1971), their results were widely used and had great intuitive appeal.

The next major attempt to classify the main settlements of England and Wales differed from its earlier counterpart in that it used data relating to 1966 (Armen, 1972). In more detail, it took in 132 variables which quantified in one way or another socio-economic status, age structure, mobility, environment, land-use and service provision. Armen also employed a methodology which produced results similar to elementary cluster analysis, but which was in fact based on a different premise, that of successive pairing of towns with a complementary partner. The advantage of this technique lay, according to Armen, in the fact that groups were not mutually exclusive and that a single town could therefore appear under a number of different headings: Portsmouth, for example, was classed both as a small port and as a day trip/vacation resort. The end product of this analysis was a typology made up of three divisions— conurbations, service centres, and manufacturing centres—which were themselves made up of eight classes and twenty-four sub-classes. Table 2.3 details this breakdown and provides examples of towns in each category. More recently, Webber and Craig (1978) have also sought to develop an urban typology using census-derived statistics for 1971. They employed variables relating to demographic characteristics, household composition, housing conditions, socio-economic structure, and various measures of employment opportunities. Furthermore, they elected to analyse these data by means of a two-stage procedure which took in both major methods of cluster analysis; namely iterative relocation, and stepwise progression. Webber and Craig combined these approaches so that iterative relocation compressed the variates into thirty groups and stepwise progression coalesced these groups into six main urban types listed in table 2.4. As with the previous studies, Webber and Craig therefore presented a typology at different levels which contained, for example, a group of rural and resort areas within which one could distinguish Rural Scottish areas from Port retirement centres. The authors concluded that their typology was of particular value because it took into account historical and regional factors as well as the more common structural variables derived from the census. They were convinced that their classification not only picked out clusters of settlements with similar characteristics, therefore, but also those with comparable backgrounds and problems. Of the studies published to date, probably the most comprehensive and valuable is that of Donnison and Soto (1980) who again worked on the 1971 census. From this, they extracted 317 variables which had some potential, but eventually narrowed these down to only 40 key variables which they felt had demonstrated some degree of statistical independence. Data on these forty criteria were gathered for 154 towns and urban areas and subjected to ten different cluster analyses. The net result was a classification into six major types with thirteen subsidiary clusters, such that the residential suburbs and New Towns were deemed to have the best quality of life, and the inner conurbations and Welsh mining towns were thought to have the worst. Donnison and Soto extended this argument further by looking at the national distribution of each type of town and delineating what they called 'new Britain' and 'traditional Britain'. The results of their typology can be found in table 2.5

The nature of Britain's ethnic settlements; 1971, 1981

Given these various attempts to categorize Britain's major cities, and the availability of unpublished tabulations of 1971 census data relating to the size of ethnic population resident in each of these cities, it should be possible to produce a parallel classification of ethnic settlements. The difficulty with such a process is that each of the five typologies of local government areas described above differ in some important respect and are therefore not directly comparable. Most consider only urban areas, but Webber and Craig include

rural areas. Some use the areal units which existed prior to the local government re-organization which produced the GLC, others use units extant in the
period between the creation of the GLC and the national reorganization of
1974, and Webber and Craig utilize the post-1974 units. And finally, each
typology selects differing criteria for segmenting the urban continuum.

Consequently, the typology of ethnic settlements which is used here does not
rely on any one of these classifications, but on a composite categorization
derived from all the previous attempts. Moreover, the typology contains only
those towns or cities which had an ethnic (i.e New Commonwealth) population
of more than 2000 people in 1971. These data were culled from unpublished
census tabulation DT 368u. In all, 45 different towns/cities were included in
the analysis, or alternatively 75 different towns/cities were included if one
were to regard each GLC borough as an entity in its own right. The New
Commonwealth population of these centres varied from 64,460 (Birmingham)
to 2020 (Ashton-under-Lyne), but between them they accounted for a total of
765,320 individuals of New Commonwealth birth, or 83 per cent of all people
of such origin in England and Wales. Ultimately, however, only 69 of the 75
towns/cities could be categorized accurately since the remaining seven had
appeared in only one or two of the urban typologies. These 69 urban areas
made up 81 per cent of the total New Commonwealth population of England
and Wales.

One is forced to say at the outset that the typology presented here is by no
means the only way in which these settlements could have been categorized. In
all classificatory procedures, no matter how apparently scientific, there is a
greater or lesser degree of subjectivity. The procedure used here was an
intuitive one in which settlements were classified according to the particular
combination of clusters or groups to which they had been allocated by previous
workers. In short, although the technique was overtly subjective it ensured that
categorization took place on a consensus of opinion.

The actual typology which resulted is detailed in table 2.6 but its main
characteristics can be outlined as follows. The continuum of ethnic settlement
is divided into four major groupings, the first of which stresses the relatively
unique nature of Greater London, and then looks for variations within this
conurbation according to the economic role, demographic characteristics, and
history of the various boroughs. The second group is made up of the other large
conurbations with their particular socio-economic opportunities and problems: these again are subdivided according to their economic base. Thirdly
there are the regional service centres with their characteristic employment
structures weighted towards non-manual occupations in either the commercial
sphere or professional and executive sectors. Finally, the fourth group contains
the major industrial centres subdivided according to their industrial allegiance
and therefore their history. One can see, therefore, that the typology blends
social, economic, and historical inputs in a way which allows a categorization of
those ethnic centres which share a common urban environment and which
therefore offer their residents similar opportunities for mobility and a similar
social context for interaction between races or ethnic groups. It is noteworthy

that in 1971 the London group contained 59 per cent of the entire New Commonwealth population being classified, with particular, and not unexpected, concentrations in the London Inner areas (30 per cent) and service centres (14 per cent). Significantly, only 7095 people of New Commonwealth (NCW) birth lived in the new industrial areas. Outside London, the other major conurbations contained an additional 15 per cent of the NCW population with the industrial centre overshadowing service centres in importance. Beyond these centres, the most significant ethnic settlements are those found in the textile (11 per cent) and engineering (10 per cent) towns, leaving the service centres (4 per cent) a relatively small role. Even at this level of analysis the conclusions bear out those of previous workers and begin to indicate the life chances which residence in Britain offers the majority of NCW people. As Donnison and Soto (1980: 114–15) point out:

Another way of describing this distribution of migrants would be to say that those born in the white Old Commonwealth and the Far East tend to go to the towns of 'new Britain' to which the native born are also moving; while those born in the black New Commonwealth tend to go to the towns of 'traditional Britain'—the Textile Towns and the Inner Conurbations—slowgrowing or declining places which the native born are leaving, where opportunities for work are generally less plentiful and less equally distributed among the social classes.

However, the typology presented above relates to conditions some twelve years ago when the British economy and many of her staple industries were in a very different condition from that of today. The intervening period has seen a sharpening rift between declining 'traditional Britain' and cushioned 'new Britain'. How has this affected the ethnic urban system?

The starting point of any such discussion must be the fact that the population of New Commonwealth or Pakistani birth (NCWP) has increased considerably between 1971 and 1981, from 0.92 million to 1.47 million. This represents a decennial increase of around 60 per cent for England and Wales. Given this fact, and what has already been said about the predominantly urban nature of this population, it seems clear that at least the scale, if not the form of the ethnic urban-system will have changed substantially over this period. This proves to be the case. Again taking the cut-off point of 2000 individuals of NCWP birth, the urban system of the ethnic population is seen to have increased by some 46 settlements to a total of 115 cities or London boroughs. This in itself represents an increase of 67 per cent in the number of settlements, or an increase of 58 per cent in the number of individuals resident within such centres.

However, probably of greater importance for our purposes here is the point that the typology of ethnic settlements which was presented in table 2.6 will no longer accommodate all those ethnic centres included in the 1981 analysis. This indicates the emergence of different types of ethnic settlements with the possibility that these might offer richer potential for social and economic advancement. Employing the same procedure as for the 1971 typology produces the ethnic hierarchy detailed in table 2.7. A point to note here is the appearance of two additional groupings of settlements; resorts are unimportant numerically or proportionally but the clusters that are titled 'Suburbs and Suburban type centres' do seem to be of some significance, particularly when one realises the preponderance of high-status or expanding centres. These naturally signify the availability of greater opportunities, and hint at the upward social mobility of at least part of the NCWP population. These preliminary indications are underscored by the rapid growth of the NCWP population in the 'New Industrial' and 'Suburban' zones of Greater London where decennial increases of 140.8 per cent and 170.1 per cent were recorded. This contrasts markedly with the slow increase of the NCWP population in Inner London (plus 22.5 per cent) and its relative lack of growth in those centres categorized as conurbations (plus 38 per cent). Interesting as the implications of these findings are, a fuller analysis of these data would be out of place here. Suffice it to say that the 1981 typology allows the identification of groups of cities which share similar histories and industrial structures, and offer similar potential to in-migrants. The purpose of such a typology, in this context at least, is to allow the more careful selection of settlements which might act as representative case studies, thereby preventing the duplication of research effort and encouraging the study of the geography of race relations as a nomothetic science rather than an idiographic art. Only when it is possible to classify situations of urban race relations (and identify the parameters which make them distinct) can we hope to progress to systematic understanding rather than particularistic description.

A Case Study of a Textile Town

Even a brief perusal of the data indicates the relatively narrow range of towns and cities which accommodate people of New Commonwealth birth. However, what the analysis does not show is the way in which academic research has focused on certain types of settlement and largely ignored others. Much of the literature, perhaps rightly in the early fact-finding stages, has been concerned with London or Birmingham but the converse of this is that the textile towns, for example, have been sadly under-researched. Why this should be, one cannot say, but it may perhaps relate to their relatively unfashionable status, the undoubted bias of the media towards the South East, or the fact that, at least from a distance, their race relations are good. None of these reasons is really adequate and case study research *is* required into these towns. When one considers that in 1971 they accommodated 11 per cent of the NCWP population and that this was only 4 per cent less than the major conurbations outside London, then they clearly merit research; especially when, as will be demonstrated later, they are rapidly growing in size and are disproportionately important for specific NCWP groups. This book therefore concerns itself with a case study of one of these settlements, Blackburn, and in the next few pages an attempt is made to validate this choice of town and describe how the textile towns differ so fundamentally from the other types of settlement in the classification.

Blackburn: the town

Blackburn, which is the subject of this study, has a total population of 142,900 and is situated about 40 kilometres north west of the Manchester conurbation and 15 kilometres east of the Central Lancashire New Town. Its development as a major city has been inextricably bound up with the advance and decline of the nation's textile industry and in particular the cotton sector of this. Natural and locational advantages ensured the rapid expansion of the town after about 1824 when local entrepreneurs began the construction of a series of large mills along the new Leeds — Liverpool canal. Blackburn soon became what Mumford (1961) terms a 'Palaeotechnic Paradise' with its fair share of railways, canals, factories, and squalor. As the population grew (see Robinson, 1982a for details of this early phase) so did the housing stock, much of which dates from the last two decades of the nineteenth century and which therefore benefited from the great Public Health Act of 1875. Consequently, the average Victorian artisan lived in what Burke (1971) labels 'tunnel-backs', a least-cost response to legislation and a dwelling-type which is almost always associated with grid-iron urban development and the close spatial proximity of residence and workplace. Blackburn, like so many other Victorian cities (see Briggs, 1968) juxtaposed this urban poverty with symbols of prosperity in the form of social buildings such as museums, town halls, and institutions, or commercial buildings such as cotton exchanges and railway stations.

However the heyday of Blackburn was not to be a particularly lengthy one and only some forty years after the metamorphosis of the town into Britain's cotton capital, the industry and the town were faced with structural unemployment and job loss on an unprecedented scale, a trend exacerbated by the Second World War. To illustrate the severity of the decline, Rodgers (1962) calculated that the contraction of the industry in Blackburn was such that two-thirds of textiles employment in the town disappeared between 1931 and 1951, whilst Smith (1969) has shown that, under the effect of the 1960 Cotton Industry Act, a further 65 per cent of jobs were lost between 1950 and 1967. Many residents of the town sought personal solutions to this economic decline by means of out-migration, and consequently each decennial census has recorded a fall in the population of between 3.3 per cent (1921–31) and 4.8 per cent (1911–21). Since migration from the town attracted the youngest and most active members of the community, the remaining population ossified socially and economically into the Cotton Culture described so well by Pearson (1976). Linked with this conservatism and concern for frugality and respectability was the progressive ageing of the population.

Paradoxically, out-migration and the progressive ageing of the workforce precipitated an unlikely crisis within the textile industry itself. The problem was a shortage of skilled labour, an issue which contemporary writers (Freeman, Rodgers, and Kinvig, 1966) considered might even threaten the very existence of what remained of the industry. Moreover, the problem was compounded by the unwillingness of married women (who formed almost 65 per cent of employees) to work on night shifts or travel more than one, or at most two, miles from their homes in order to get to work. In retrospect, it seems that it was this vacuum within the employment structure which encouraged the settlement of the first Asian families in the town during the late 1950s and early 1960s, encouraged ·also by the relative cheapness of obsolescent terraced houses.

The Asians thus arrived in Blackburn at a time of crisis; the staple textile industry had been savaged by foreign competition (much of it ironically from India) and it no longer provided a viable economic base; the housing stock had remained largely untouched since its creation in the 1880s and was therefore rapidly becoming time-expired; the town was suffering a haemorrhage of young and educated people who left behind them a physically and psychologically aged population; and as a direct product of these other trends, Cotton Culture was in disarray, collapsing under the onslaught of unemployment, modernity, and permissiveness. Pearson (1976) argues that the perceived collapse of a traditional way of life was instrumental in generating racist sentiments and racist violence in North East Lancashire as the 'Pakis' became increasingly to be seen as the cause of this collapse rather than as coincidental bystanders. Even if this is overstating the case, one has to admit that it was crucial in creating an atmosphere which other trends brought to fruition.

Like so many other perceptions, though, the common stereotype of Blackburn which many people hold in the 1980s is long overdue for updating. Much has taken place since the crisis-laden early 1960s and the city has undergone a considerable physical metamorphosis even if it still houses the same people with the same outlook.

It is another paradox of the textile industry that its very decline created conditions conducive to the growth of industries which would replace it. The stable and manually-skilled workforce, and the cheap premises which became available as successive mills closed, both attracted the development of new employment sectors, most notably engineering, electronics, clothing, footwear, carpet manufacture, and paper-making. By the mid 1960s then, textiles was no longer the main employer, as the production of TV tubes, car radios, alarm systems, and other electrical components pushed engineering and electronics to the forefront. One must, of course, not give the impression that this remodelling of the local economy happened spontaneously, since both the government (through Development Area status) and the local council were very active indeed in creating a conducive economic climate for the growth or relocation of firms.

The second major area where the town has cast off its previous image has been the urban environment itself, where extensive and continuing programmes of inner area redevelopment have turned Blackburn from being an introspective textile town into an exciting and dynamic regional centre. Much of this new image has been generated through redevelopment of the town's Central Business District and its ancillary functions. In the same way that the Victorian era saw conspicuous consumption on public buildings so have the 1960s, 1970s, and 1980s. Much of the central area of the town has been demolished to make way for extensive and adventurous shopping centres, malls, and department stores. These shops and facilities now attract a clientele from a large surrounding hinterland, and Blackburn is vying for importance with the traditional administrative capital of Lancashire, Preston.

Thirdly, by the 1960s Blackburn was also faced with a housing crisis. This again sounds paradoxical in view of what has been said about out-migration and the declining size of the town's population, but Blackburn's crisis was not concerned with the *number* of dwelling units available to the local population but the quality of these units. As has been noted, Blackburn expanded rapidly in a very brief period at the close of the nineteenth century when a large proportion of its tunnel-backs were constructed. Seventy years later, this period of expansion had created a legacy of housing, much of which had become obsolete at about the same time, both in relation to the amenities on offer and to the environmental quality of the neighbourhoods within which it was situated. Consequently, the major issue faced by Blackburn has not been the extent to which the housing stock needs expanding but how the authority can most effectively modernize the stock through refurbishment or clearance. The full details of how the authority has faced this challenge are contained in Robinson (1982a: 10–26), but in practice the planning and housing departments have used a combination of clearance, modernization, and council building; the combination changing in line with national policy directives.

In the sphere of public housing, the authority has benefited from each of the main initiatives, with major estates being completed during 1924 (eight estates as a result of the 1919 and 1923 Housing Acts), between 1946 and 1953 (eleven estates and estate extensions), between 1961 and 1966 (ten new or extended estates), between 1967 and 1971 (nine developments), and between 1971 and 1979 (a further ten projects). The spatial pattern of these developments can be seen in figure 2.1. Each of these phases within the town reflects changing priorities at a national level, and therefore different arrangements for funding. By 1981, then, the local authority (as constituted by the 1974 re-organization) contained some 14,780 council dwellings (Chartered Institute of Public Finance and Accountancy, 1982) of which around 12,000 are located within the old county borough. Table 2.8 provides a breakdown of these properties by type, size, and age and indicates the relatively youthful nature of the stock and what is now seen as the unfortunate number of flats within this total. Whilst on this subject, one should note that Blackburn's planning ideals for public housing have also shifted in line with national fashions. The older peripheral estates are large and contain single family dwellings in the form of terraces or semi-detached houses. The aura of these estates clearly evokes Ebenezer Howard's (1902) Garden Suburbs. Those estates found in the middle ring (built during the 1960s) were also conceived on a grand scale and tend to be multi-storey complexes or collections of tower blocks. In contrast, the more recent developments near the centre of the town are planned on a much smaller scale and contain high density and high amenity houses in terraces or pairs. These properties deliberately ape the design of the older tunnel-backs and therefore seem both more appropriate and more familiar.

It would be wrong to think that the local authority has set about improving the housing stock merely by council building. The 1961 census indicated that

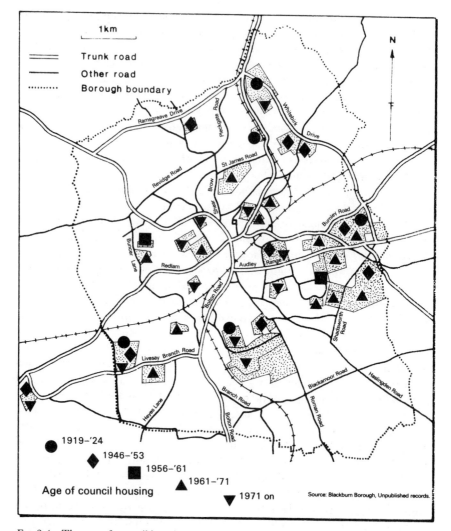

FIG 2.1: The age of council housing in Blackburn

the percentage of households without sole use of a cold water tap was 0.08; the proportion of households without a bath was 40 per cent; and that approximately 4 per cent of the population did not have sole use of a flushing toilet. In the light of these statistics, and encouraged by the 1956 Housing Subsidies Act, the Borough drew up a Clearance Programme in 1963. This was to be phased over a twenty-year period and took in 10,500 properties or approximately 30 per cent of the total housing stock. By 1966 the Borough had already demolished over 6000 of these houses and by 1979 (Blackburn Borough, 1979) the Planning Department was able to comment that any further clearance in the 1980s would be on a smaller and more sensitive scale.

As was the case with the town's council building programmes, the policies

which have been employed to upgrade the quality of housing through refurbishment have also changed in line with national policies and objectives. Thus
the late 1960s saw a major shift from area clearance to the concept of gradual
renewal (see Gibson and Langstaff, 1977 for the reasons behind this). This new
concept was embodied in the 1969 Housing Act which spawned the General
Improvement Area, or GIA, which has been defined by Roberts and Gunby
(1973) and criticized by Duncan (1974) and Short and Bassett (1978).
Blackburn's response to this legislation was immediate and consisted of the
declaration of the town's first GIA in 1970, a project which by 1979 had been
joined by 29 other private sector, and 12 municipal sector declarations. By
1979 these policy instruments had produced upgrading which had been completed in 3400 owner occupied houses, whilst a further 4000 grants had been
approved.

In 1974, the Department of the Environment responded to criticism by
extending the criteria which defined areas worthy of grant-aided improvement. The result was the Housing Action Area, or HAA, which despite its
greater generosity and generality did not escape widespread criticism (see
Lawless, 1977; Freeman, 1977; Short and Bassett, 1979; Jones, 1980). Even
though 'in many respects, the new provisions are unhelpful to . . . Blackburn.
The social criteria for HAA declaration . . . are hardly fulfilled outside the
main conurbation centres' (Blackburn Borough, 1975), Blackburn *has* made
use of the HAA as another vehicle to modernize its housing stock. The first
HAA in the town was duly declared in 1976 and it was expected that this would
be followed by a further five HAAs by the end of 1981.

The net effect of these individual changes has been to revitalize and
reorientate the town away from King Cotton, and its environmental and economic legacies, towards a new future. The town now has a broader and more
secure economic base, a far better infrastructure, and a much improved
housing stock located in environmentally richer neighbourhoods. In short, in a
period of only twenty years Blackburn has changed from being a laggard which
relied upon its past glories, to being an innovator which has created its own
future. Having said this, though, one cannot escape the fact that much has not
changed. The mills still line the canal even though they may now be
hypermarkets. The back-to-backs still rise up the hills in rows away from the
spinal Leeds – Liverpool canal. The symbols of a previous era still stand in the
form of the Old Town Hall, King George's Hall, and the railway station. And
above all else, most of the inhabitants are still tainted with cotton culture and
the attitudes and outlook which go with it. In this sense, despite the changing
economic and physical landscape, Blackburn can still be categorized as a textile
town even if engineering is now the more important employer.

Consequently there are very good grounds, as the various typologies suggest, for regarding Blackburn as a representative example of the textile towns
which are located on both flanks of the Pennines. And, similarly, there are
grounds for thinking that much of what occurs in Blackburn with regard to race
relations may equally well occur in Oldham, Bolton, Huddersfield, and
Bradford. As a result, we shall argue that any conclusions which derive from

this study of Blackburn do have implications outside the town and do form a representative study of this type of town. Moreover, previous sections not only show what problems and issues have faced Blackburn during the last one hundred years, but also begin to define what makes towns like Blackburn different from Coventry, Leicester, or Liverpool. This is a theme to which Chapter Eight returns when it discusses how structural and attitudinal constraints bear upon new in-migrants. However, before it is appropriate to describe how Asian migrants came to find themselves in a town such as Blackburn, it is necessary to pursue a little further what it is that makes the textile towns such a unique group of settlements. Some indication of this can be gathered by describing the characteristics of Blackburn's socio-residential structure and comparing this with national data.

The socio-residential structure of Blackburn

For the purpose of this investigation it seems appropriate to work with a classification of social areas or social 'types' which is not specific to one data set or urban area. Consequently, the national classification of Enumeration Districts (EDs) generated by the Centre for Environmental Studies was used. This was essentially an extension of the urban typology of Webber and Craig (1978), described above, to individual EDs rather than individual towns. It allows one to say for example that ED 152 in Blackburn shares similar characteristics with ED 143 in Swansea, or ED 4 in Leeds. It thus provides national comparability. For the ED analysis, the same 40 variables were used as inputs but in this case they produced sixty clusters or eight broad families. The results of the procedure as applied to the EDs of the old County Borough of Blackburn were that seven of the eight possible families were represented in the town. The missing family was, not surprisingly, that which described rural areas. The seven categories which remained could be described or summarized as follows:

Family 2:   areas of established high status and elderly population
Family 3:   new owner-occupied estates of high status and young age structure
Family 4:   areas of older terraced housing and mixed tenure
Family 5:   areas of extensive public housing
Family 6:   areas of extensive public housing and acute social stress
Family 7:   areas of low status multi-occupied housing with serious social stress
Family 8:   areas of high status rented housing, students and other single people

In Blackburn, although seven of the eight families are represented, there is a marked dominance of EDs in family 4, followed by lesser concentrations in families 5, 2, and 7 in respective order. This pattern is not disturbed by considering the percentage of the population (rather than the number of EDs) in each of the families. As table 2.9 illustrates, over half the population is resident in areas of older terraced housing with mixed tenure. A comparison of the percentage of the population in each family at a national and local level reveals the discrepancies between Blackburn's socio-spatial composition and that of the national average and therefore quantifies the uniqueness of the textile towns. As expected, this reveals that the town has an over-representation of population in areas of older housing which do (the minority) or do not (the majority) exhibit signs of social stress. Significant under-representation occurs in families 3 (new owner-occupied estates), 8 (high status rented), 6 (public housing and social stress), and 5 (areas of extensive council housing). The spatial distribution of these families within the old Blackburn Borough is shown in figure 2.2 which immediately demonstrates the juxtaposition of high status areas (families 2 and 3) to the west, north west, and north of the town—on higher land and near municipal parks—and areas of municipal housing (families 5 and 6) which are located to the south west, south, east, and north

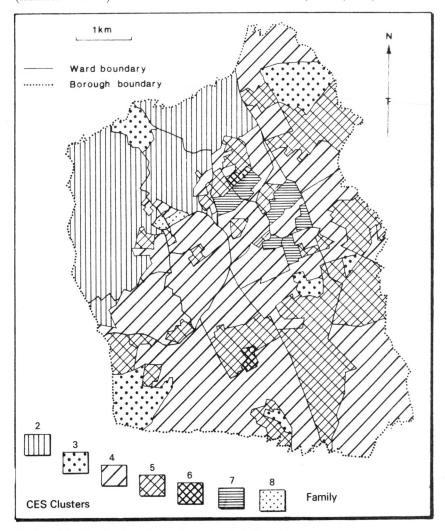

FIG 2.2: The distribution of CES cluster analysis families

east of the town. Enclosed within this perimeter and following the lines of rivers and railways are the areas of terraced housing (families 4 and 7).

A general picture derived from study of the broadly defined families suffers from the fact that each family contains a range of differing clusters which exhibit varying characteristics. An analysis at cluster level is more difficult but is more rewarding. Altogether, thirty-five clusters were represented amongst the EDs of Blackburn; a figure which represents approximately 58 per cent of the range of possibilities. Table 2.10 compares the percentage of Blackburn's population found in these clusters with the percentage of the national population. It is not practicable to provide either a verbal description or a profile of every one of these thirty-five clusters. As an alternative the upper quartile of clusters was extracted so that the characteristics of those clusters which contain the largest percentage of Blackburn's population could be described in detail. These nine clusters account for over 65 per cent of the town's population. They are described in Appendix I, and comprise a very accurate description of the social structure of Blackburn in 1971. Taken in conjunction with figure 2.2 they also allow us to construct an idealized diagram of Blackburn's socio-spatial structure (figure 2.3). This makes an interesting comparison with Mann's (1965) hypothesized English industrial city and will also provide a useful *aide-mémoire* to the context in which the processes and patterns outlined in the following chapters have taken place.

The Textile Towns and Asian Settlements

Previous sections have described a typology of ethnic settlement in Britain based on those cities which had a resident New Commonwealth population which exceeded 2000 in 1971 or 1981, and have defined within this a reasonably homogenous group of towns which has been labelled the textile towns. It has also been argued that Blackburn represents a good example of this group of towns and that by studying this one particular example we may cast some light on the situation in the others. However, in order to provide a broad over-view, the discussion so far has concerned itself with the census-defined NCWP population. It will become apparent in later chapters of this book that one of its central arguments is that such gross aggregations of the coloured populations are neither conceptually nor empirically sound, and that such insensitivity plays a central role in blunting policy.

This section therefore narrows the focus of the analysis a little and considers the importance of the textile towns for the Asian population alone. It is at this point, then, that further analysis of the Caribbean, African, or Mediterranean New Commonwealth populations is excluded. Some idea of why this might be fruitful can be gained from the fact that correlation of the number of West Indians found in the 23 largest ethnic settlements in 1981 (excluding the GLC and Birmingham) with the number of people of Asian birth or descent in these same settlements in 1981 produced a coefficient of only 0.18. Clearly then, whilst the distributions of these two groups do overlap there are obviously fundamental differences which demand that the two groups are analysed separately.

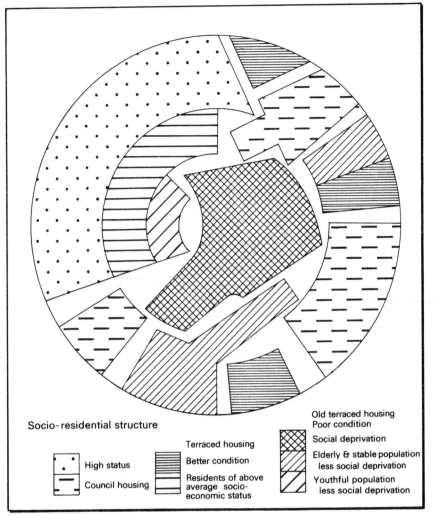

Socio-residential structure

| | | Old terraced housing Poor condition |

Terraced housing
Better condition

Social deprivation

High status

Residents of above average socio-economic status

Elderly & stable population less social deprivation

Council housing

Youthful population less social deprivation

FIG 2.3: Idealized model of the socio-residential structure of Blackburn

The starting point is a consideration of the ethnic hierarchy purely for the Asian population. Table 2.11 lists the twenty largest Asian settlements in England and Wales both in 1971 and 1981. Attention is drawn to the notes beneath this table since they contain some important provisos to any conclusions derived from it. Comparison of this table with table 2.1 reveals some interesting differences. Whilst, not unexpectedly, Birmingham and the GLC come out as the leading centres in both the NC and Asian hierarchies, beneath these conurbations rank placings begin to shift. Significant losers of rank are Nottingham (down 8 places), Sheffield (down 5), Manchester (down 4), and Bristol and Reading, both of which drop from the ranking altogether. Conversely, gains in rank are recorded by Bolton (up 6 places), Blackburn (up 5), Slough (up 5) and by Rochdale and Preston, both of which enter the table. At this level of analysis then, it would appear that the textile towns are of greater importance to the Asian group than to the other NC peoples. A similar conclusion is derived from comparing the 1981 NC and Asian rank orders. Blackburn scores the largest gain in rank (up 7), whilst Huddersfield, Leeds, Bolton, Oldham, Preston, and Rochdale all have higher ranks within the Asian hierarchy than they do within the broader NC hierarchy. This emphasizes the claim that a study of the textile towns is of especial importance for an understanding of the Asian population and therefore why research on Blackburn is of significance despite the town's relatively small population in comparison to places such as the GLC or Birmingham. The rise in importance of Blackburn as an Asian settlement (a gain of three places in the period 1971–81) also validates the decision to choose this town in preference to others such as Bolton, Oldham, or Rochdale.

One can, of course, be rather more rigorous about drawing conclusions, given the typology of ethnic settlements drawn up earlier in the chapter. Table 2.12 therefore provides data on the distribution of the Asian population over the different categories of settlement in both 1971 and 1981. The respective NC figures are also presented to aid comparison, as are location quotients which quantify the degree of Asian over-representation (above 1) or under-representation (below 1) in each urban type relative to the whole NC population. It is clear that the Asian population was over-represented to a considerable degree in both 1971 and 1981 in the textile and textile/service towns. Moreover the degree of over-representation remained fairly steady between the censuses or even increased. Beyond this immediate concern with the textile centres, it is also noteworthy that Asians are over-represented in the new industrial areas of London (although the extent of this declined in the decade 1971–81), the mixed residential suburbs of the capital, and the engineering and heavy engineering centres. Persistent under-representation is manifested within the GLC's inner areas and its service centres, commercial service centres, and engineering and commercial cities. During the 1970s, significant changes in status were recorded in the professional and administrative centres where the Asian population failed to grow at the same rate as the whole NC population, and in the new industrial and exclusive suburbs of London where the Asian population declined in percentage and relative terms. Clearly, whether these trends indicate worsening economic prospects for Britain's Asian population is a separate research issue which ought to be pursued with some urgency. However, for the purpose of this study the crucial issue is that, when taken alone, the textile towns are the second most important category within the distribution of the Asian population and, when they are combined with the two textile/service towns, they form the most important single category. According to the typology presented here, more than one in every five Asians lives within a textile or textile/service centre. It is obvious therefore that research into a representative example of this type of ethnic settlement would be of both value and significance.

Summary

This chapter has argued that research into the geography of ethnic settlement in the UK needs to become more systematic and more rigorous. The simple duplication of local community studies picked at random by researchers working in isolation is unlikely to yield any great depth or breadth of understanding. As a result, a typology of ethnic settlement in British cities has been presented in this chapter as an essential first step towards encouraging a more scientific and rational approach. Such a typology also emphasizes the importance of research into the textile towns, of which Blackburn was selected as a representative example. Finally, the account outlined those features of Blackburn's social and economic make up which are considered to be characteristic of the textile towns and which therefore make them different from say Sparkbrook or Handsworth in Birmingham, and how this character becomes translated into potential life-chances for Asian and white residents.

CHAPTER 3

The Migration of Asians to Britain

This chapter is concerned with the general characteristics of South Asian and East African Asian migration to the UK. It describes the scale, timing, and causes of the two streams of international migration as an essential prerequisite to the more detailed and localized study of those Asians who eventually settled in Blackburn, and whose migration histories are discussed in greater detail in chapter 7. As with the previous chapter this material serves to illustrate how the study of one particular group of individuals nests into broader international and national circumstances.

The analysis of Asian migration to the UK and the subsequent settlement of those groups in urban ethnic areas has to be approached by way of a stage-by-stage chronology since Asian communities have existed in British cities for over 120 years. Indeed Asians were not unknown as performers in seventeenth century Britain and as domestic servants for the nation's upper classes during the following century (see Fryer, 1984 for an excellent account). Several chronologies have been suggested but the most comprehensive is that proposed by the Ballards (1977). This is used as a framework for the description of South Asian settlements but would be inappropriate for the later East African Asian movement. Consequently, a different approach is employed in that section of the chapter.

South Asian Migration

The early pioneer period

Discounting the settlement of servants and performers, Asian migration to Britain really began early in the twentieth century with the arrival of single males who had migrated spontaneously either for adventure (Aurora, 1967) or for economic gain. Those in the latter category were often associated with the shipping industry either as stokers (the Mirpuris) or as galley-hands (the Sylhetis). It is now thought that they only gained access to these jobs because of the shortage of indigenous labour brought about by the Great War.

Occasionally, when they docked at British ports the Indian seamen would jump ship and seek alternative employment. Some became labourers in heavy industry, whilst others established themselves in self-employment either as tea-shop proprietors (the Sylhetis: Little, 1948) or as door-to-door pedlars selling cheap clothing (the Sikhs: Faux, 1980). Often, however, migrants did not stop in Britain for more than three or four years, although on their return to India many were replaced overseas by close relatives in a system of rotating transience.

The small numbers involved in the pioneer phase and the impermanence of settlement prevented the development of uniquely ethnic areas in British cities. Even in 1949, for example, there were still fewer than 100 Indians in Birmingham. As a result, the pioneers occupied scattered accommodation in the twilight zone of Britain's cities where they followed the Poles and other East Europeans in a loose ecological succession (see Griffith, 1960 for examples).

The lodging house era

Causes and characteristics. In time, both push and pull factors united to encourage mass migration. High rural population densities, pressure on the land, fragmentation of landholdings because of the laws of inheritance, the Partition, and local phenomena such as the construction of the Mangla Dam all combined to create unusually strong push factors. Conversely, the buoyancy of the British economy and associated opportunities for economic and social gain, and the tradition of overseas migration all acted as powerful enabling or pull factors. However whilst most authors agree that this combination of factors coincided to produce mass migration, there is no agreement about the relative importance of push and pull factors. Some authors argue that emigration was forced upon Asians as a reaction to downward social mobility (e.g. Aurora, 1967) whilst others insist that emigration provided a voluntary avenue to rapid economic and social advancement (e.g. John, 1969). This disagreement is not without significance since the causes of emigration are likely to have determined the characteristics of the personnel involved. However, Pettigrew (1972) has sought a *rapprochement*. She argues that the cultural ethos which underpins the sending society is more important than the specific motives behind migration. The sending society is characterized by a search for improved status or honour (*izzet*) which is equated with material wealth. However, because of the laws of inheritance, material wealth and therefore status is but transitory. Each individual within a new generation is consequently motivated to aspire. Without this motivation, status is threatened because of the inevitable subdivision of inheritance. Individuals who are migrating in order to improve economic status and those migrating to stave off a loss of *izzet* are thus opposite poles on an intervening continuum. The former group is seeking to increase status prior to, and in anticipation of subdivision, whilst the latter group is acting to retrieve status lost through recent subdivision.

Pettigrew's notion that Asian migration to Britain was controlled by both push *and* pull forces depending upon the specific position of individual migrants is borne out by more quantitative analyses. Whilst Peach (1965; 1967; 1968; 1979) was able to demonstrate the overwhelming importance of economic 'pull' factors for West Indian migration, a similar analysis of South Asian migration undertaken by Robinson (1980a) produced more ambiguous results. Like Peach, Robinson used Asian immigration figures derived from Home Office sources rather than the alternative data from the International Passenger Survey as recommended by Jones (1981a; 1981b). Published sources point to the clear superiority of the former over the latter (Runnymede

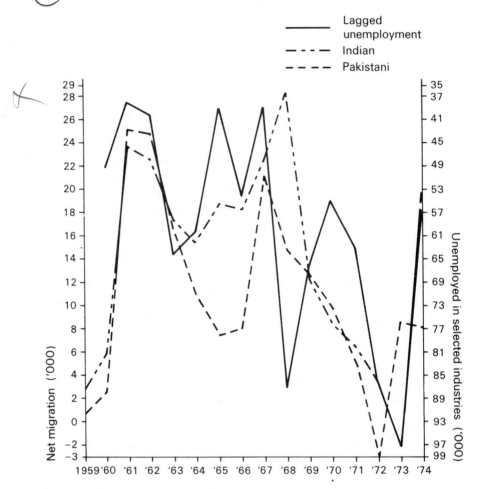

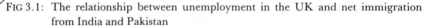

FIG 3.1: The relationship between unemployment in the UK and net immigration from India and Pakistan

Trust, 1980; Peach, 1981a; 1981b). Robinson's correlation of these data with a measure of labour demand in relevant industries suggested that the relationship between these two variables was by no means as strong for South Asians as it was for West Indians (see figure 3.1). Clearly, economic pull factors were not the only components stimulating migration. Whilst some migrants would have the resources to await an opportune time to come to Britain, others would have to take whatever opportunities existed. Migration would thus take place regardless of the state of Britain's economy.

Macro-level patterns. Once Asians had gained entry to the UK, many of their settlement patterns were determined purely by economic, or pull factors. Peach (1966) argued that West Indians avoided areas of low demand for labour and moved instead to regions where labour demand was high and unsatisfied.

Jones (1978) considers that this is also true of the Asian population. He suggests that the pattern of Asian settlement contains two separate elements, both related to labour shortages. Firstly, settlement took place in areas where white labour was scarce because of the pace of economic expansion e.g. Greater London, Birmingham, and their respective satellites. Secondly, settlement also occurred in those areas where there was a labour shortage in certain industries because of the poor conditions of employment e.g. the textile industry of Manchester, Leeds, and proximate towns.

Census data allow a more detailed analysis of patterns of settlement during the lodging house era. Table 3.1 lists the distribution of both main birthplace

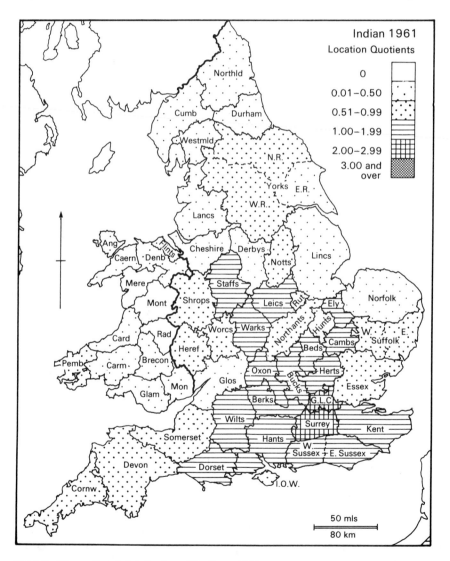

FIG 3.2: The national distribution of Indians, 1961

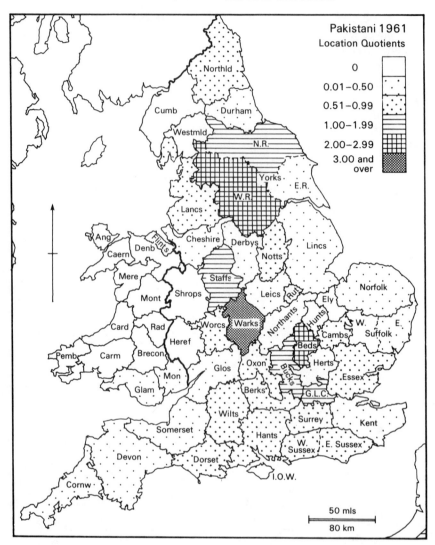

FIG 3.3: The national distribution of Pakistanis, 1961

groups by economic sub-region for 1961. It demonstrates the primarily urban,
or conurban distribution of both groups such that over a half of Indians, and
two-thirds of Pakistanis were resident in the six main conurbations. Figures 3.2
and 3.3 allow a more detailed description of these patterns since they map
Location Quotients for both groups at the finer county level. Indians were
over-represented in seventeen counties, all of which were found in the more
affluent South East and Midlands. London, Middlesex, and Surrey appear to
be the central foci of the distribution whilst the other counties of over-
representation form a triangle around this core with vertices in Dorset,
Staffordshire, and Kent. This implies a conscious regional location orientated

towards areas of high status and high labour demand. The former may well
result from the inclusion in census data of Indian-born whites, which Peach
and Winchester (1974) estimate may have formed as much as 70 per cent of the
total Indian population of the UK in 1961. In contrast, and perhaps a reflection
of the absence of Pakistani-born whites, the Pakistani population was more
clustered. Only six counties evidenced relative over-representation and these
were industrial counties such as London, Bedfordshire, Warwickshire, and the
East and West Ridings.

  Both patterns therefore clearly demonstrate the way in which Asians acted as
an industrial replacement population in areas of rapid industrial growth (the
South East) and in areas of industrial decline and out-migration (West
Yorkshire). The same conclusion can also be drawn from Robinson's (1980a)
study of the relationship between demand for labour within one town
(Blackburn) and the net growth of the Asian population there (figure 3.4).

Ethnic areas and housing. The lodging house era saw the rapid immigration of
single, male Asians to a small number of towns or cities within the UK. Once
settled, several factors encouraged new Asian migrants to cluster in existing
locations within urban areas. Firstly, the impermanence and economic orientation of migration were reflected in the continued demand for privately rented
accommodation. Such accommodation was only available cheaply within the

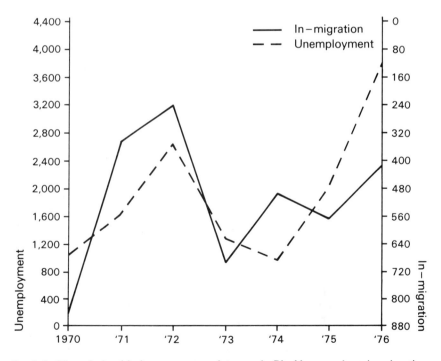

FIG 3.4: The relationship between unemployment in Blackburn and net immigration
     of Asians to the town

zone of transition. Secondly, chain migration and sponsored passages ensured that new arrivals were funnelled into existing areas of settlement near kin and friends. The support of the *biraderi* (brotherhood or fraternity) and community would be at their most effective within such reception areas. Thirdly, the desire to maintain social encapsulation (see chapter Five) encouraged voluntary clustering: propinquity allowed regular interaction between fellow-migrants and facilitated the construction of pseudo-traditional social networks and community institutions. Increasingly then, Asians became concentrated in inner-city areas, firstly in the ubiquitous lodging house owned by whites and latterly in the all-male Asian household.

Jones (1967) was one of the first researchers to study the segregation of immigrant communities during the lodging house phase. His concern was to explain and describe the form and intensity of residential segregation in Birmingham. Jones found that the Asian population was not concentrated into the worst housing in Birmingham or into the very centre of the city. Rather, they were segregated into the larger Edwardian and Victorian terraces of the city's middle ring. These houses, which were available cheaply because of their short leases, were well-suited to conversion into lodging houses and were purchased by groups of individuals using short-term loans from friends or relatives. They were then sub-divided and let out, either to kin or friends, or as an economic enterprise, to strangers. The resulting lifestyles which developed were described by Rex and Moore (1967). Other authors have commented upon similar patterns of housing preference and spatial clustering e.g. Husain (1975) in Nottingham, Jones and McEvoy (1974) in Huddersfield, Kearsley and Srivastava (1974) in Glasgow, Richmond (1973) in Bristol and Winchester (1974) in Coventry.

Family reunion

Causes and characteristics. Family reunion was characterized by the arrival in the UK of dependants who served to complete the joint household. This entailed the migration of the wives, sons, daughters, and to a lesser extent parents of the 1960s target migrants.

Jeffery (1973) suggests three reasons why the lodging house era gave way to that of family reunion during the late 1960s and early 1970s; firstly, a belief that a wife could only execute her duties properly when living in the same household as her husband; secondly, the need to have the father present in order to discipline children; and thirdly, the fear that unaccompanied residence in Britain might encourage the husband to be unfaithful. This fear was heightened by a common belief that English society was morally lax (see chapter 5). The Ballards (1977) suggest the additional reason that the increasing complexity and completeness of the social networks of Asian males in Britain encouraged competition for status solely within the arena of the British settlement. As hospitality and the entertainment of guests was a crucial factor in re-establishing *izzet* within the British arena, the presence of womenfolk became socially vital: the house had to be kept clean and tidy to impress guests and food had to be prepared.

As a result of these four factors the migration of dependants gathered momentum during the late 1960s and early 1970s. Table 3.2 shows the shifting sex balance of Asian immigration to the UK during this period, and Jones (1978) provides census data to show the stabilizing balance of the sexes within the Asian minority in the UK. Smith (1977) provides further data on this topic, summarized in table 3.3. These demonstrate how the male dominance of the lodging house era was reduced, but not eradicated, by family reunion. It is also worthy of note that family reunion took place despite increasing administrative obstacles (Moore and Wallace, 1975) and despite the fear of cultural contamination of wives and daughters (Wilson, 1978).

Macro-level patterns. Comparison of 1971 and 1961 census data reveals the impact of the family reunion phase on the spatial distribution of the two groups. Table 3.4 provides a breakdown by economic sub-region and reveals that the Indian-born population had become more concentrated in the English conurbations (56.8 per cent against 52.3 per cent) than it had been in 1961, whilst the Pakistani-born population became less concentrated in these centres (63.7 per cent against 68.8 per cent). The former trend could be explained by the declining relative importance of dispersed 'white Indians' but the latter seems to evidence dispersal to secondary settlements. In greater detail, the Indian population became more concentrated in the West Midlands, West Yorkshire, South East Lancashire, and the outer Metropolitan area. Peripheral regions such as Wales, the South West, and the North all lost population in relative terms. All the major conurbations, except one, also lost their relative importance for the Pakistani minority. Major growth areas for this group were the North West region in general and the South Lancashire conurbation in particular, and the Metropolitan areas of the South East.

Figures 3.5 and 3.6 provide a more detailed picture using county level Location Quotients. Comparison of figure 3.5 with figure 3.2 makes clear the process of Indian centralization between 1961 and 1971. Only nine counties had an Indian population which was over-represented in 1971 and a large number of counties had changed from being areas of slight under-representation to being areas of marked under-representation. The pattern of Indian concentration crystallized around London and the Midlands although Kent, Buckinghamshire, and Bedfordshire retained their relative positions. Comparison of figure 3.6 and 3.3 reveals the parallel process of Pakistani decentralization. Five counties attracted sufficient Pakistanis to become areas of over-representation for the first time: Lancashire, Worcestershire, Berkshire, Huntingdonshire, and Outer London. By 1971 the Pakistani population was more evenly distributed throughout the 'industrial coffin' from the Pennines to London.

Jones (1978) has discussed the impact which family reunion had upon the pattern of urban, rather than regional, settlement. He identified two clusters of towns which had experienced 'very fast' or 'fast' growth as a result of the arrival of dependants. The Lancashire textile towns of Bolton, Blackburn, and Rochdale formed one cluster, whilst the other was made up of the engineering

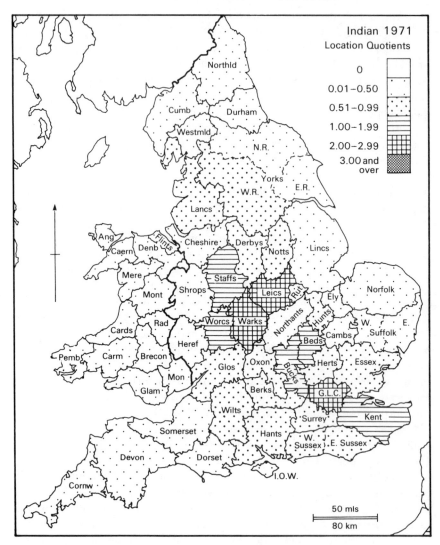

FIG 3.5: The national distribution of Indians, 1971

towns of the East and West Midlands such as Leicester, Coventry, and
Walsall. Conversely, because of their demographic maturity, the Asian popu-
lations of other important centres stagnated, e.g. Manchester, Nottingham,
Sheffield, and Bristol.

Ethnic areas and housing. The creation of Asian nuclear, or extended, families
precipitated a change in the housing requirements of the minority. The lodging
houses were replaced by cheap, unimproved inner area terraces. These houses
and their facilities have been described by a number of authors: Robinson
(1979a) for Blackburn; Jeffery (1976) for Bristol; Goodall (1966) and Duncan

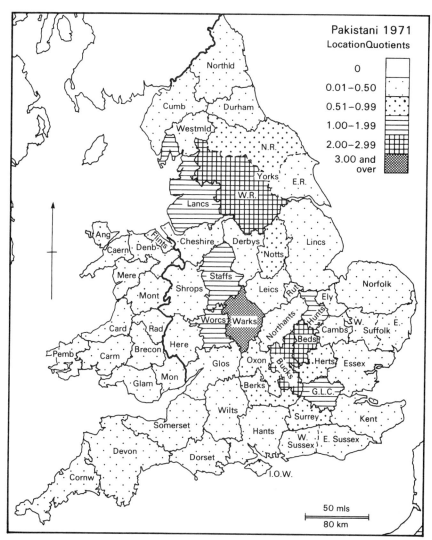

FIG 3.6: The national distribution of Pakistanis, 1971

(1977) for Huddersfield; Werbner (1979) for Manchester; and Dhanjal (1976) for Southall.

Since many of the new secondary centres of Asian settlement were only formed during the family reunion stage, the arrival of dependants encouraged growth *in situ*. In longer-established centres, though, family reunion forced a physical relocation of the minority out from the twilight zone. In certain cases this took the form of incremental expansion into contiguous areas of owner-occupied property e.g. Birmingham (Jones, 1970; Woods, 1979), Bradford (Khan, 1977), London (Lee, 1973), and Rochdale (Anwar, 1979). In other cases, the movement into single-family dwellings stimulated the development of newly-pioneered ethnic space away from pre-existing clusters. Werbner's (1979) account of the relocation of Pakistanis in Manchester is an excellent example of this trend.

Maturity

Government statistics now suggest that the population of Asian ethnicity is one of the more rapidly growing sectors of society. Estimates published by the Immigrant Statistics Unit of the Office of Population Censuses and Surveys (OPCS) (1977) indicate that the Asian minority has grown from forming under 40 per cent of the total coloured population in 1971 to around 45 per cent in 1976. More recent data from the 1981 Labour Force Survey (1983) suggest that this figure is now nearer 49 per cent. In numerical terms, this growth represents a change from 546,000 people in 1971, to 796,000 in 1976, and to 1,054,000 in 1981. The Immigrant Statistics Unit (1979) estimates that the population of wholly Asian ethnic origin might well rise to between 1.25 and 1.5 million by 1991.

Whilst it is clear from these data that the Asian population is growing rapidly, aggregate figures do not reveal the source of this growth or whether it is evenly distributed between the Indian, Pakistani, and East African populations. Table 3.5 answers these questions, and indicates how, between 1971 and 1976, the most rapidly growing group were the East African Asians (as a result of refugee immigration) followed by Pakistanis and Indians. For both of these last two groups, natural increase had become more important than net immigration. These figures suggest that, with the exception of isolated refugee movements, the Asian population is now entering a period of self-generative growth, and family formation or expansion. These trends indicate a movement towards demographic maturity.

Analysis of other data confirms this shift towards maturity although it is important to note that different sub-groups are progressing towards this goal at different rates. The number of births to Indian mothers in Britain has remained constant between 1973 and 1981 despite a rise in the number of Indian women (Thompson, 1982). The Total Period Fertility Rate for women of Indian, Pakistani, or Bangladeshi birth has fallen from 5.7 in 1970 to 4.5 in 1975 (Iliffe, 1978). The percentage of all three main birthplace groups aged over 45 years has increased between 1971 and 1981. And the sex ratios of all three groups became considerably more balanced over the same period.

In sum then, the Asian minority seems to be entering a period of maturity and stability in which certain groups engage in the family building postponed during the earlier phases of the movement, and other groups manifest characteristics convergent with those of the West Indian and white populations.

Macro-level pattern. 1981 census data quantify the extent of any changes in spatial distribution which have accompanied maturity. Table 3.6 provides a breakdown of distribution by economic sub-region. A comparison of this table with table 3.4 is not strictly accurate because of the changes in boundaries which occurred in 1974 but it does isolate general trends. The first point to note is that both South Asian groups appear to have become more concentrated in the English conurbations. This still remains true when the newly-created Tyneside conurbation is excluded from calculations. The Indian population contained in the six conurbations rose from 57 per cent in 1971 to 61 per cent in 1981. Similarly, Pakistani concentration increased from 64 per cent to 67 per cent. However, aggregate conclusions such as these are misleading. They imply a process of recentralization throughout the country whereas, in practice, this was not the case. Of the six main conurbations, only Greater London increased its share of both the Indian and Pakistani population. The other

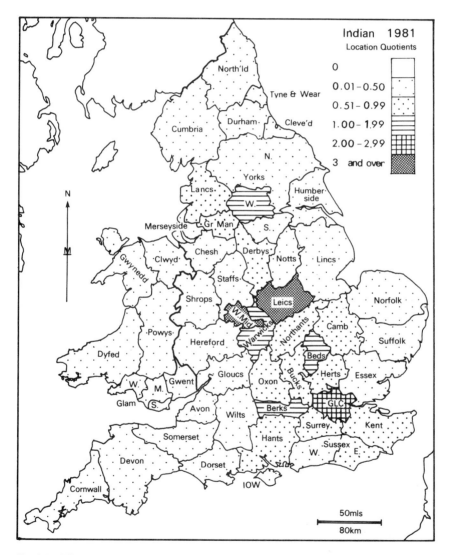

FIG 3.7: The national distribution of Indians, 1981

conurbations either remained stable or lost population in a relative sense. Outside the conurbations, small percentage gains were recorded by non-conurban Lancashire (Pakistanis only), the East Midlands (both groups), the outer South East (Pakistanis only), and Wales (Pakistanis only).

Figures 3.7 and 3.8 provide county level location quotients. Again, comparison of 1981 and 1971 data is made difficult by the 1974 local government reorganization. However OPCS republished 1971 census data for post-1974 counties so direct comparison is possible. The maps reveal the basic stability of both distributions over the period 1971 to 1981. For the Indian population, there were only slight changes in county location quotients with the exception

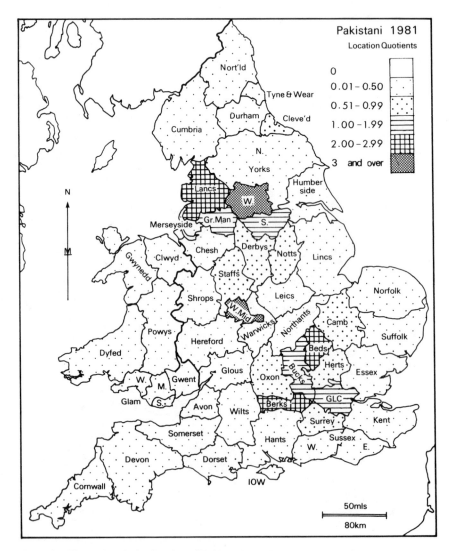

FIG 3.8: The national distribution of Pakistanis, 1981

of Leicestershire which for the first time overtook the West Midlands as the area of maximum over-representation (2.57 in 1971 and 3.21 in 1981). Other counties retained their relative positions. For the Pakistani population the rank order of the leading eight counties remained unchanged but the actual value of their location quotients rose, indicating population growth in existing areas of over-representation. Overall though, the phase of maturity is characterized by a stable population distributed in an almost static pattern.

Ethnic areas and housing. As a response both to demographic expansion and to the early stages of social mobility (Field *et al.*, 1981), maturity has encouraged the development of a fourth stage in the process of settlement. This involves decentralization to better quality properties found at a distance from the initial 'ports-of-entry'. Such properties are owner occupied and are often purchased by means of a building society mortgage. They are more modern, more expensive, larger, and better equipped.

As yet geographical evidence for the suburbanization of elements within the Asian population is still limited. Dhanjal (1976) provides an impressionistic account of Asian movement from the traditional core areas of Southall to semi-detached properties in neighbouring Harrow, Hounslow, and Hillingdon. Husain (1975) comments on the movement of Asians from the centre of Nottingham to peripheral estates in the Urban District. Finally, Werbner (1979) documents decentralization in South Manchester towards Chorlton-cum-Hardy, Withington, and Stockport.

Both Werbner, and Nowikowski and Ward (1978) stress that there is a good deal of heterogeneity amongst those Asians who are resident in the suburbs. Nowikowski and Ward identify three main groups with differing backgrounds: educated professionals who had come to Britain for access to higher education and who had therefore begun life in this country in rented student 'digs'; cosmopolitan businessmen who had come to Britain to expand the activities of their companies and who immediately took up residence in the suburbs; and emergent traders who had worked their way up from unskilled manual employment and were now self-employed retailers, wholesalers, or manufacturers. This group would have arrived in the suburbs by way of step-wise migration from core areas.

East African Asian Migration

Few of the generalizations which apply to South Asian migration and migrants also apply to their counterparts from East Africa. This largely results from three factors: firstly that East African Asians are not primary migrants in the sense that they have moved directly from their ancestral homelands to Britain; secondly that they came from a different colonial social structure from that which existed in India; and thirdly that they were refugees rather than voluntary migrants. In combination these points go a considerable way towards explaining many of the characteristic features of the migration stream and its constituent members, but they are also worthy of individual discussion.

Asian settlement in East Africa

One of the main myths which surrounds the East African Asians is that they were the descendants of coolies introduced by the British to build the East Africa Railway. In practice, Gujurati settlements existed in Africa as early as the thirteenth century and Delf (1963: 1) even goes so far as to suggest that 'the Indians were from the earliest days the masters of finance, the bankers and money-changers and money-lenders'. This obvious business expertise endeared the Indian merchants to the Arabs who had gained an early hold over the eastern coastal areas, and who offered tolerance, patronage, and immunity in exchange for successful economic management. The strength of those links was such that, by 1860, the Indian merchants are said to have controlled almost all the Zanzibari trade (Swinerton *et al.*, 1975). This very success was instrumental in gaining the Hindu traders a reputation as being clannish, unscrupulous, and exploitative, and they were also thought to be mere 'birds of passage' without any commitment to Zanzibar (Mangat, 1969).

The decline of Arab power in East Africa saw the Indians adjusting to new conditions and new overlords, but these circumstances did not pose any serious threat to their privileged intermediate economic and social status. British rule did, however, introduce new economic opportunities and new Asian groups to exploit these. The need for labour to construct and man the new East Africa Railway led to the introduction of 32,000 Punjabi (Sikh and Muslim) labourers under six-year contracts of indenture (Tinker, 1977). The scale and rapidity of this migration was responsible for the emergent stereotype that 'Indian' was synonymous with 'uneducated coolie'. The facts belie this generalization since it ignored the dominant Gujurati traders and it ignored the fact that of 32,000 migrant coolies, only 6724 stayed on after the expiry of their contract (*New Community*, 1972). The remainder died, were invalided home, or returned to India of their own free will. New economic opportunities also encouraged the voluntary migration to East Africa of Gujurati clerks, traders, artisans, and a few professionals. Many of these did not take up farming as the British authorities had anticipated but instead turned to trading as *dukawallas*, or shopkeepers. In this role they became middlemen who sold soap and textiles to Africans in exchange for their cotton, coffee, or ivory. They also became the chief source of credit for the African peasantry, and naturally attracted all the opprobrium which money-lenders usually receive. One cannot, however, question the success of Asian commerce. Before the Second World War Asians controlled over 90 per cent of the total trade of Uganda whereas in Tanganyika they were responsible for 80 per cent of the cotton trade, 50 per cent of imports, 60 per cent of exports, and 80 per cent of transport services.

Their intermediate economic and social position has been further strengthened since the last war by the increasing emphasis which the Asians have placed upon education and social mobility (see Rattansi and Abdulla, 1970). In Uganda and Tanganyika the Indians were the driving force behind the establishment of the first schools, initially to transmit their ancestral culture but latterly as a route to the professions. By the early 1960s Asians had gained considerable access to both the professions and the Civil Service. In 1962,

Cable (1969) reported that 23 per cent of economically active male Asians in Kenya were involved in the professions or management, 42 per cent in clerical and sales posts, 26 per cent as skilled craftsmen, and only 5 per cent as semi- or unskilled workers. Ghai (1970) provides comparable figures for Tanzania where 35 per cent of workers were in executive, managerial, or administrative posts, 25 per cent in skilled manual, 20 per cent in secretarial and clerical, and 15 per cent in professional and technical jobs. This new position of inter-mediate economic status based upon education was reflected in a range of other characteristics such as qualifications, lifestyles, and income levels (see *New Community*, 1972). As Cable (1969: 3) put it, 'undoubtedly the Asians consti-tuted under the colonial system a privileged minority which did nothing to divest itself of its relatively superior status. Most were prepared to accept the restrictions of racial segregation that attended it'.

However, the relative success of the Asian minority in terms of wealth or position within the social structure was never matched by the extent of their social acceptance. Whites either viewed them as a necessary pariah group which had to be tolerated or as direct competitors for political and economic power. Black Africans viewed them as a privileged minority which had access to scarce resources and jobs, and which was exploitative by nature and only a transient part of society. These negative evaluations were not helped by the social exclusiveness of Asian society and the inability of that minority to present a united and effective public front because of internal communal divisions. Bharati (1970) identified six distinct Asian sub-communities in East Africa whilst Ghai and Ghai (1970), Delf (1963), and Swinerton *et al.*, (1975) dis-cussed the social, political, and demographic ramifications of this.

The coming of independence, Africanization, and emigration

The independence of the three East African colonies from British rule heralded the beginning of a new era of adjustment for the Asians. Initially little changed, but the Asians were largely sceptical of the African's ability to rule the newly independent states. This was manifested in three ways: many Asians, and in particular Kenyans, began to transfer their wealth out of the continent; there was a rapid decline in immigration and in certain cases, panic emigration e.g. the Goans from Kampala (Kuper, 1975); and many Asians chose not to adopt the citizenship of the newly independent state but to retain their British citizen-ship. As *New Community* (1972: 409) put it 'the advantages and disadvantages of each nationality were shrewdly calculated, like the credit and debit sides in their account books.'

The question of citizenship was merely a prelude to the beginnings of Afri-canization, although it did emphasize to many the relatively uncommitted attitude of the Asian minority. Partly because of this, but more importantly because of the issues of remittances and Asian domination of the economy, similar policies of Africanization began to appear in many of the new states. Kenya was the first to adopt such policies when she openly encouraged Africans to gain entry to the Civil Service at the expense of Asians, to the point where, in

1967, all non-citizens were removed from public employment. This was followed in the same year by the Immigration Act, the Trades Licensing Act, and the Transport Licensing Act. In 1969, Milton Obote followed Kenyatta's example, and Ugandanization began to parallel Kenyanization. The impact upon the Asian minority was obvious and relatively immediate. In 1965, 5000 Kenyan Asians left for Britain, in 1966 this rose to 6000, in 1967 to 12,000, and in January and February of 1968 a further 12,000 left. In Uganda, between 1962 and 1969, one in every twenty-five Asians migrated to Britain. However this was clearly a situation which was not acceptable to the British government who promptly and summarily introduced the ill-conceived Immigration Act (UK) on 27 February 1968.

The legislation did little to prevent a recurrence of the problem although it took a little over four years for the next crisis to materialize in the shape of the expulsion of the Ugandan Asians. In 1968 Obote set about rectifying what he identified as Asian dominance of commerce in Uganda. In January 1971 though, Obote was overthrown by the army and replaced by General Amin. At first the Asian community considered this to be fortunate, but the virulently anti-Asian attitude of the army soon generated misgivings. These were strengthened when Amin ordered an Asian census nine months later and when he convened an Asian conference at the end of 1971. At this conference Amin levelled many stereotyped criticisms at the Asian minority and demanded that leaders of the community should reply to these allegations. In particular he questioned the social exclusivity of Asian society in Uganda and the relative absence of exogamy to Africans. The Asian leaders replied in a Memorandum published on 8 December (see Twaddle, 1975) which seemed to silence Amin, for no further action was taken. This relative quiescence encouraged Asians to think that the worst was over and, according to *New Community* (1972), few actually left the country as a direct result of Amin's early actions. That is not to say that emigration was absent. Between 1969 and 1971 over 24,000 Asians had left Uganda, largely to come to Britain, although others went to the United States and Canada whilst many of the older generation retired to India. The fact that many of these emigrants left Uganda with the maximum amount of money (£5000 initially, but £2500 later) aggravated the situation for those who remained, for they were forced to bear criticism stemming from the perception of them as uncommitted economic transients. Finally on 4 August, Amin made public his intentions. All non-citizen Asians were to leave Uganda within 90 days, an order which was later temporarily expanded to include *all* Asians regardless of citizenship.

It is clear from this brief description of events that analysis of East African Asian migration to the UK cannot satisfactorily take place within the framework of concepts such as 'push' and 'pull' forces. The Kenyan, Ugandan, and latterly Malawian Asians were above all else political or economic refugees rather than voluntary migrants seeking greater social status.

Characteristics of the migration

Because of the compressed time scale over which the East African Asians migrated to Britain, it is more difficult, but not impossible, to recognize discrete temporal stages in the migration.

The majority of those East Africans who entered Britain prior to 1967 were young, single men sent from Kenya by their families either for access to higher education or to assess the suitability of the UK as a destination for the remainder of the family, should the need arise. Because of the cost of travel, these pioneers were drawn mostly from the wealthiest Asian families in East Africa. However, as policies of Africanization began to take effect in Kenya and Uganda, the pioneers were joined by the small shopkeepers and clerks who had lost their livelihoods. According to Cable (1969), these new arrivals were neither the most, nor the least, progressive members of their communities. Tinker (1977) suggested that they also occupied an intermediate economic position. The new arrivals also differed from the pioneers in their demographic characteristics. Rather then being single men, the 'anticipatory refugees' (Kunz, 1973) consisted of family units with a roughly equal sex balance and a broader spread of ages. It was noteworthy, though, that the migration still contained few aged people since this group tended to return to India on their departure from East Africa. Both of these demographic features are illustrated by table 3.7 which provides an age and sex breakdown for all three Asian groups.

By the close of the 'voluntary' phase of Asian movement from East Africa, the Immigrant Statistics Unit of OPCS (1975) estimated that there were approximately 68,000 East African Asians in Britain, and that the minority formed approximately 12.5 per cent of the UK Asian population and 5.2 per cent of the total coloured population.

The migration changed character for a third time with the involuntary expulsion of the Ugandan Asians. Such a movement could not, by nature, be selective and the refugees who arrived in Britain therefore belonged to a wide range of social classes and the full array of age groups. Again family units, and in particular extended family units, were the norm. This last feature has tended to give the East African Asian minority in the UK a unique character since the Indian, and more particularly the Pakistani groups contain few elderly dependants.

The 1981 Labour Force Survey (1983) illustrates how the arrival of the Ugandan Asians transformed the size of Britain's East African Asian population. By 1981, the population of East African Asians by birth had risen to some 155,000, a growth of 87,000 in only ten years. Brown (1984) indicates that the East African population has both a larger proportion of extended households than any of the other coloured groups and a larger percentage of pensioner-only households. These data clearly reflect the more balanced nature of the East African refugee stream.

Macro-level patterns. The difficulty in any analysis of the spatial distribution of East African Asians in Britain is that there is no census category which relates solely to this group. The 'East African Commonwealth' category contains not only Asians but black Africans and whites. Data are therefore not 'pure'.

Whilst this had become less of a problem by 1981 (by which time Asians formed 75 per cent of this census category), it was particularly acute in 1961. The 1961 census data are therefore not analysed for that reason.

The 1971 census indicates the nascent patterns of settlement developing from voluntary migration. Only five counties exhibited over-representation, these being Leicestershire (LQ of 5.62), Greater London (2.78), Warwickshire (1.48), Bedfordshire (1.41), and Buckinghamshire (1.36). Comparison of figure 3.9 with figure 3.5 reveals that the East African Asian population had a very similar pattern of distribution in 1971 to that of the Indian population. Differences centred upon the greater concentration of East African Asians in

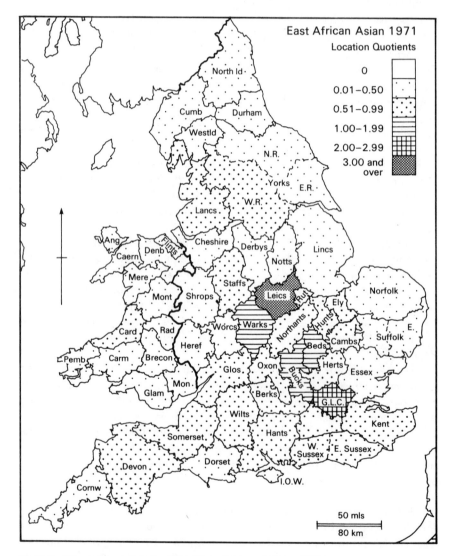

FIG 3.9: The national distribution of East African Asians, 1971

the East Midlands whereas Indians were more concentrated in the West Midlands.

It is impossible to discuss the macro-level distribution of East Africans during the 1970s without recourse to a consideration of the resettlement policies instituted by the government during the Ugandan Asian crisis. The government of the day was faced with the problem of resettling 28,608 (Uganda Resettlement Board, 1974) Ugandan Asians who were forced to take at least short-term refuge in the UK. These people had been given insufficient time to sell their capital assets, had not been granted compensation in lieu of these, and had only been allowed to take out limited amounts of cash. In short, the Ugandan refugees had very limited resources with which to facilitate their own resettlement. Set against the need to provide immediate and generous aid to compensate for this, was the attitude of the electorate towards further coloured immigration and immigrants. At the time of Amin's decree of expulsion, but prior to the entry of the Ugandan Asians, over 57 per cent of a Gallup sample professed themselves doubtful as to whether Ugandan Asians should be admitted to the UK (Kohler, 1973). The government thus had to balance its moral and historical obligations against electoral support.

The immediate response of the government was to establish the Uganda Resettlement Board (URB). The URB presided over the creation of sixteen 'Transit Camps', in which the refugees had their immediate needs of food, warmth, shelter, and companionship met by voluntary groups such as the WRVS and the Red Cross. However, it is not really these reception policies which are of interest. The parallel programme of resettlement had a more direct bearing upon the evolution of patterns of spatial distribution. In formulating this programme the government had to bear in mind clear signs of opposition to the resettlement of refugees in particular cities (Bristow, 1976). As a consequence, the URB soon adopted a policy of dispersal rather than concentration of refugees. Whilst it could be argued that this not only minimized the social cost of resettlement to the indigenous population but also gave the best possible opportunities to the refugees, Ward (1973), at least, considers that it was the former reason which prompted the adoption of this goal not the latter.

The dispersal policy was made public in the form of 'red' or 'no go' areas, and 'green' or 'go' areas. The selection of these was made ostensibly on the grounds of the presence of pressure on housing, education, or social resources but the existence of glaring anomalies suggests otherwise. Glasgow, a city with an enormous housing problem but without a large ethnic community, was declared green (Dines, 1973). Leicester, which suffered only limited housing problems but which had a large pre-existing Asian community became a red area. Once a city was declared red, the URB did its best to discourage refugees from settling there, whilst trying to stimulate offers of accommodation from local authorities which had been declared green. At first, offers were few in number since local authorities did not wish voluntarily to burden their own ratepayers with the cost of resettling the Ugandan Asians. Later, however, once central government offered compensation through the Rate Support

Grant and Section 11 payments, the offers became more numerous.

By 1 January 1973 the URB had amassed a motley collection of property amounting to some 2100 dwellings scattered throughout 340 different local authorities. The problem remained of matching these properties to the 7775 Asians who were resident in URB camps on the same date. A major difficulty in this process concerned the geographical distribution of these offers, which was controlled to a greater extent by generosity and the demographic structure of local authorities, than by the needs of the Ugandan Asians or available employment opportunities. Many local authorities were only able to offer accommodation because the indigenous population had already decided that the local economy offered few opportunities, and had therefore left the area. It was unlikely that the Ugandans could have succeeded where the local population had failed. In addition many Ugandan Asians only possessed knowledge of Leicester and London and were unwilling to move far from these places, let alone to Scotland or the South West. Thus, as table 3.8 reveals, there proved to be a considerable mismatch between offers of accommodation and the desire to take up these offers. Despite this, Swinerton *et al.* (1975) argue that the relocation achievements of the URB were most impressive and that many Asian families who would naturally have gravitated towards areas of housing stress were steered towards areas without these problems, a view which received some support from the URB's (1974) final report (see table 3.9). Other commentators are less sanguine about the achievements of the URB. Bristow (1976) for example notes that although 38 per cent of those Ugandans who passed through the hands of the URB *were* eventually found places in green areas, this still left 62 per cent who were not. For the majority, the 'normal' working of the housing market prevented any dispersal to green areas and, by default, they were consigned to the deprived and depressed red areas.

Bristow and Adams (1977) in a follow-up study of Ugandan Asian housing found little evidence to support a national redistribution of the refugees after resettlement. The residential mobility which was present related to movements within the twilight zone rather than long-distance relocation. Consequently, the 1981 pattern of distribution can be regarded in large measure as a product of resettlement policies combined with the voluntary patterns of the earlier, but less numerous, anticipatory migrants. The national pattern of distribution is shown in table 3.10 which should be compared with table 3.6. Such a comparison reveals that the East African population is less concentrated in the six main English conurbations than either the Indian or Pakistani groups (59.8 per cent against 61 per cent and 67 per cent respectively). The converse of this is that the East Africans appear to be more heavily concentrated in certain regions of the country. Their regional distribution is in some ways similar to that of the Indians, with several important exceptions. The East Africans are found less in the northern regions (the North, Yorkshire and Humberside, and the North West) and less in the West Midlands. Yet they are concentrated more in the East Midlands and Greater London. Data at the regional level, then, suggest that the government's policy of dispersal succeeded in producing a distribution which was more concentrated than those which developed naturally for Indians and Pakistanis. This is borne out by figure 3.10 which illustrates the bi-polar nature of settlement. Such a conclusion is not without parallels in Britain's history of refugee programmes (Robinson, 1985b).

Ethnic areas and housing. For those Ugandan Asians resettled by the government in green areas, tenure patterns will have directly reflected the resettlement programme's emphasis on the provision of council housing. Bristow and Adam's (1977) data confirm that council renting is the dominant type of tenure for this group. Of their interviewees who were resident in green areas without sizeable pre-existing Asian populations, 73 per cent were council tenants, 22 per cent were visiting, 3 per cent were owner occupiers and 2 per cent were

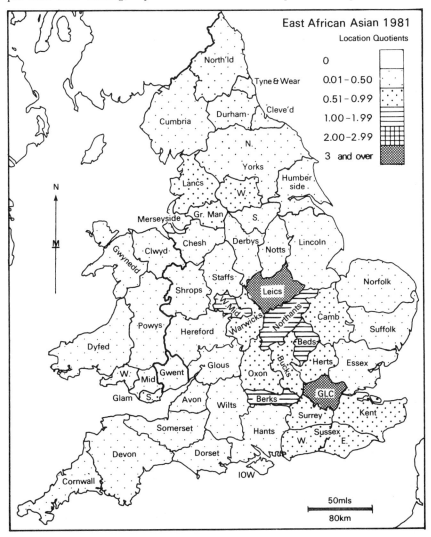

FIG 3.10: The national distribution of East African Asians, 1981

staying with relatives. Bristow and Adams concluded that those Ugandan Asians who had been fortunate enough to be resettled in a green area would most probably remain there, that their tenure patterns would remain fairly stable, but that there might be some movement into owner occupation in parallel to upward social mobility.

Those refugees who chose, or were forced, to take accommodation in red areas or green areas with large pre-existing ethnic populations were not so fortunate. They had relied upon social contacts to help them find homes and were therefore placed in the same position as any other low income family without permanent employment or substantial collateral. In such a position it was natural to turn either to private renting or to relatives for temporary or even long-term shelter. Bristow and Adams thus found that 47 per cent of their respondents were renting property from private bodies, 47 per cent were staying with kin, 4 per cent were owner occupiers and 3 households were council tenants. A longitudinal analysis of changes in tenure revealed a shift towards owner occupation and council renting, but private renting decreased only marginally.

Bristow and Adams concluded that the government and URB had created two distinct groups of Ugandan Asians: one suffered all the disadvantages of inner-city living from which they were unable to escape; the other benefited from the government's ability to overrule the normal working of the housing market and were consequently in a relatively advantaged position. Work on the housing of this latter group includes Bristow (1979), Kumar (1973), and McCart (1973).

However, in view of the fact that the Ugandan Asians are not the only East African group in Britain it is important that their experiences and character-istics are not emphasised to the exclusion of other groups. Two main data sources allow a broader analysis of the entire minority: these are the 1974 and 1982 PSI/PEP studies of Smith (1977) and Brown (1984). The former was undertaken shortly after the Ugandan Asian crisis and therefore provides a useful 'snapshot' of the state of development of the minority at this point in time. Smith found that East Africans had a distinctly lower level of owner occupation and a higher level of private renting. They were housed at higher densities than other Asian groups and were often the tenants of Indian land-lords. Smith puts their relative disadvantage down to the 'time factor', i.e. the recency of their migration. Whilst Smith therefore dwells on disadvantage, a closer analysis of the data suggests other conclusions: East Africans had already exceeded white and West Indian levels of owner occupation by 1974; 19 per cent of African Asians lived in a house with an amenity score of less than 9 (where 0 is the minimum and 12 the maximum) compared to 25 per cent of Indians and 44 per cent of Pakistanis; 31 per cent of African Asians did *not* have exclusive use of a bath, hot water, and an inside WC against 35 per cent of Indians and 57 per cent of Pakistanis; and lastly, of those owner occupiers who were interviewed, 7 per cent of African Asians were sharing a property com-pared with 10 per cent of Indians and 14 per cent of Pakistanis. Clearly then, despite the 'time factor' the African Asians had already made rapid strides towards good housing and could hardly be viewed as an especially disadvantaged group.

Brown's (1984) data seem to confirm this notion of the East Africans as a progressive and dynamic group. By 1982, African Asians had reduced their private renting to levels comparable with other Asian groups and lower than those of the white or West Indian population. They had gained greater access to council housing. They were more likely to live in a detached or semi-detached property than any other coloured group. They had a smaller percentage living in pre-1919 houses than any other coloured group. They scored well on all indices of overcrowding. And they had above average percentages of households with gardens, central heating, and telephones. In short, the East Africans could not be regarded as one of the more disadvantaged ethnic groups despite the time factor. This is an important point to which we return in subsequent chapters.

The presence of a government policy for the resettlement of at least some of the East African Asians in Britain meant that the patterns produced by the residence of this group are many and varied. The variety is further enhanced by the differing housing demands of contrasting social status groups within the minority. As was noted above, the richer refugees who arrived at the beginning of the exodus made their way to the GLC where they settled in the north western suburbs (*New Community*, 1972). In particular, Brent proved to be attractive (7.3 per cent of the first 2101 heads of households), as did Barnet (5.1 per cent), Ealing (3.8 per cent), Harrow (3.2 per cent), Newham (3.1 per cent), and Hounslow (2.6 per cent). *New Community's* correspondent noted, in addition, that even those families who were going to boroughs where there were already concentrations of immigrants were avoiding these ethnic areas and settling in the more affluent neighbourhoods. Clearly, disadvantaged ethnic areas did not result from the settlement of this sector of the minority. The same could also be said of those Ugandan Asians dispersed to council housing in green areas since their patterns of settlement would be determined by the availability of property at the time of the crisis.

For the poorer East Africans and those Ugandans who made their way to red areas the situation would be different. As the example of Leicester indicates, the late arrival of the East Africans meant that they were entering a fully developed and institutionally complete ethnic settlement with its own internal order and housing shortage. Consequently, after a period in rented accommodation Kenyan and Ugandan Asians by-passed the Victorian terraces of the Narborough Road and the Highfields area and moved directly into newly pioneered ethnic territory in the Belgrave area north of the city centre. These houses were both larger and better than those of the pre-existing ethnic core. Over a short period of time, therefore, the East Africans pioneered new and better neighbourhoods and created their own ethnic territory from which they were keen to exclude other Asians. Robinson (1980b) reports similar pioneering of new ethnic space and opportunities by the East Africans in Blackburn. For the majority of East Africans, then, residence is within ethnic settlements but within discrete neighbourhoods of these and within better quality housing than that of their South Asian counterparts.

Summary

This chapter has analysed the differing characteristics of South Asian and East African migration to Britain. A chronological approach was adopted for South Asian migration. For each of the four main stages data and discussion were provided on causes, characteristics and spatial outcomes. In addition, all of the sections contained a general account of the changing demand for housing which each of the stages generated. Above all, though, the section on South Asian migration stressed the overriding importance of economic motives in stimulating Indian and Pakistani migration and also in determining the type and location of housing to which South Asians were able, and chose, to gain access.

In contrast, the account of East African Asian migration stresses not voluntary or economic causes but the fact that the movement was compulsory, politically induced, and unplanned. The account also highlights a second important difference between East African and South Asian migrants to Britain. The former came from an intermediate stratum within colonial society and they, therefore, had higher expectations than their South Asian counterparts who were largely drawn from the lower strata. This has clear implications for the roles which the two groups might adopt in Britain.

CHAPTER 4

Structural Forces

Previous chapters have outlined the environmental and urban context into which international migration has channelled Asian migrants to Britain in general, and Blackburn in particular. They have also looked at that migration itself in order to place the Blackburn migrants in both a national and international context. The next two chapters are concerned with less concrete phenomena since they address the issue of the forces which bear upon the average South Asian or East African Asian during his residence in Britain. This present chapter focuses upon those forces which emanate from the receiving society and which result from the structure of that society. The next chapter turns to the forces which migrants carry with them from the sending society and which are determined by the culture which prevails in that society.

Critical Elements Determining Social Position

There is a strong tradition in western industrialized societies, such as our own, of the dissolution of class divisions, an increasing emphasis upon individual achievement, and the existence of a state of democratic equality (Ward, 1975). Rex and Tomlinson (1979) argue that this has come about as a result of what is, in effect, a class stalemate. Although class conflict has yielded significant welfare rights, political participation, and the right to collective bargaining, these achievements have paradoxically served to stabilize the capitalist system rather than undermine it. The working class realizes that it has made significant gains but that more substantial change is unlikely. It consequently adopts a defensive posture designed to maintain solidarity through the instruments of the trade unions and the Labour party. Above all, it takes as its prime purpose the securing of those gains which have already been achieved. Again, paradoxically, whilst the class as a whole retains its commitment to incremental gains through solidarity, the individuals within that class also believe in individual social mobility through capitalism. The middle class, in contrast, retains a continuity of action and more importantly a continuity of (somewhat diminished) control over the functioning and norms of society. Rex (1970) thus suggests that British society is now characterized by cultural unity, value consensus, and normative sanctions aimed at the maintenance of full employment and the welfare state. Although this consensus is occasionally punctuated by conflict and overt bargaining, the overall economic order is accepted. Status consequently displaces class as the key social differentiator and individuals strive to take advantage of the opportunities which exist for social mobility into the middle or higher strata of society (Goldthorpe, 1980).

Unlike class relations, status is less easily conceived in terms of discrete and mutually exclusive units, and is commonly conceptualized as being a continuous gradient (Cox, 1948). The three key variables which allow an individual to place himself, or others, on this continuum are occupation, education, and residence (Rex, 1970). However an individual's status is continually under threat because of the need to interact in daily life with people of dissimilar status. Mechanisms are thus required to prevent the collapse of the status hierarchy. These can take the form of the conscious or subconscious reinforcement of status identity through membership of associations, or the creation of social distance through manipulating the form and content of out-group interaction. Social distance can be created between individuals of differing status by ensuring that relations either remain what Mitchell (1966) has termed 'categorical' or 'structural'. The combination of groupings resulting from objective positioning in relation to the market and perceived social status, produces a fine subdivision of society exemplified by Rex's (1961) differentiation of the 'clean poor' and the 'lower-lower class', and Ward's (1975) division of workers into 'ultra-respectables', 'respectables', and 'roughs'.

The final element to consider when discussing the structure of any given society is the unequal distribution of access to political power, an element which Dahrendorf (1959), at least, considers to be paramount. Whilst this may be isolated as a separate constituent in its own right, it must clearly be seen as part of those already mentioned. Location in the economic structure is therefore at least a partly determining factor in creating the political power structure. Similarly, status groupings are not without a power input since each group seeks to impose its values on all others.

The cross-cutting affiliations produced by differential access to political power and differential location with regard to economic production and other markets consequently generate a complex multi-faceted stratification system in any society, and in this Britain is no exception. As a result, even without significant cultural differences such as those found in the United States it becomes difficult to think in concepts such as 'assimilation to the host society' or 'adoption of the norms and values of the indigenous population'. Clearly the host society or indigenous population is made up of an amalgam of sub-groups with varying attitudes, values, aspirations, and political and economic power. This is a point taken up in chapter 6.

Attitudes to Coloured People

One area in which Rex's argument about value consensus seems to have relevance is the issue of attitudes towards immigration and the settlement of coloured people in Britain. Society may well be a republic of different status groups who are seeking to gain or protect different privileges but little evidence exists to suggest that attitudes to coloured people vary greatly between these groups. Given this consistency, it is possible to make generalizations, not only about the causes of such attitudes but also about the way in which they are translated into behaviour.

The true causes of prejudice are still not understood although it is becoming clear that no single cause can explain the existence of prejudice in all individuals. Some authors may rely upon economic or structural explanations (e.g. Cox, 1948) whilst others propose psychological ones (e.g. Adorno, 1964) but the truth seems to be nearer to Allport's (1954) assertion that these approaches are merely opposite extremes on a continuum of causes and that their emphasis varies for different individuals. However, in the British context there is a significant body of academic opinion which argues that an essential prerequisite to any discussion of the causes of prejudice is a study of colonialism and the social contact which it produced (Bagley, 1970; Lawrence, 1974; Rex, 1970). Rex was one of the first to argue in a comprehensive manner that in order to understand contemporary race relations it is necessary to adopt an historical approach. He suggested that explanations of current attitudes are rooted in the context within which early contacts took place between the two groups concerned. For Asians and West Indians, this context was one of open conflict during British overseas colonial expansion. The outcome of this initial contact was, of course, the establishment of British colonial domination through the political, economic, and military subjugation of indigenous peoples. The latter were incorporated into a new and imposed societal structure in which they were rarely seen as equals. Their position in this new hierarchy labelled them as conquered and inferior people. Rex goes on to argue that, subsequently, if either group thinks of the other they will be reminded of these early contacts and the societal positions which resulted from them. However, these early experiences are strengthened by several other factors. Firstly, if the group categorized as being inferior is definable by a physical characteristic, then that characteristic itself becomes a mark of inferiority, and it also facilitates more effective and more persistent maltreatment. Secondly, stereotypes may be strengthened by the economic structure: political and military domination might allow an ethnic division of labour in which the oppressed group is forced to work for the dominant group either on plantations or as indentured labour. Slavery is the most extreme example of such circumstances. Eventually, therefore, the conquered indigenous peoples are allocated a role not only as military, political, and social inferiors but also as economic inferiors who come to be regarded as raw materials in the process of production. This will be exacerbated where members of the metropolitan society and the dominated society are forced to interact by economic or social imperatives, since the former group is likely to emphasize stereotypes in order to prevent its own de-statusing.

So much for the colonial society, but these processes only become of importance to the metropolitan society if attitudes are transferred there and given widespread currency. A brief consideration of the situation reveals, however, that this is the case. Rex (1970) argues that those individuals who transfer knowledge about the colonial society back to the metropolitan population are unlikely to do so in an unbiased manner. The ex-colonial officials were the very people responsible for employing force in the colonies and are likely to see the indigenous population as 'savages', 'terrorists', or 'neo-slaves'. The politician or student who has visited the colony will have done so with the preconceived notion that the country is inferior, backward, or underdeveloped. He is likely to reiterate these ideas on his return. Even the families of colonial officials are likely to disseminate stereotypes since their only contact with local people was with servants or tradesmen. In short, the attitudes of the colonial rulers are transferred intact to the metropolitan society where they form the major part of the information available about the colony and its inhabitants. In time, such attitudes and ideas are internalized in the culture and institutions of the metropolitan society. Although few English people had been to India or even met an Indian:

> Generations of them grew up in the period when this country ruled vast areas of the world and hundreds of million of black people. And they learned, as a result, not only that it was good to be British but that it was unfortunate and a sign of inferiority to be black. Geography lessons at school were a constant reminder of this for large areas of the world's map were coloured red. History recorded the triumphs of the British army but the treachery of the Indian Mutiny and the barbarity of the 'Black Hole of Calcutta'. On Empire Day children sang 'Land of Hope and Glory' and 'Rule Britannia' and were reminded of Britain's supremacy. And in Sunday School they learned of the daring exploits of white British missionaries who, without a thought for their own lives, took light into the gloom of darkest Africa and the hearts of the black heathen natives who lived there. . . . they learned that black people were different, that we were their masters and teachers, and that they were naturally subordinate to us. (Lawrence, 1974: 55.)

Even before their arrival in substantial numbers in Britain, then, coloured people were perceived as being different and inferior (see Lapiere, 1928).

As was made clear in chapter 3, when coloured immigrants first arrived in Britain they were allocated those housing tenures and types which were least in demand from the indigenous population, and in a similar fashion they were given those jobs which were either too unpleasant or too arduous for British workers. Nevertheless there was a considerable spatial, residential, and occupational overlap between the new arrivals and the established population. The crucial question was what would be the attitude of the established population when they were forced to share space, houses, and jobs with the immigrants. Would the presence of common problems encourage class consciousness? Clearly, the preconceived attitudes towards coloured colonials ensured that few members of the lowest stratum of metropolitan society identified with the new arrivals. To admit that you shared similar problems and a similar position to an inferior and stereotyped black neighbour or workmate also necessitated an admisson that you yourself might be inferior. As a result the metropolitan working class became as keen, if not keener, than other classes to prove that black immigrants were not only different but very definitely inferior. The stereotype of the 'vanquished peasant' or the 'slave' was thus embellished to include such characteristics as 'dirty', 'promiscuous', 'troublemaker', 'overcrowder', and 'ghetto-dweller'. As a result, the indigenous working class could feel distanced from and socially superior to their black neighbours or colleagues. De-statusing had been averted again. Rex and Tomlinson (1979) add to this largely social explanation the economic factors which produced a similar outcome. They argue that there was an international division of labour between metropolitan and colonial society and that this ensured that it was in

the interests of middle and lower metropolitan classes to 'unite in the exploitation of, and in defence against, any threat from segments or groups within or
from colonial society . . . The relevant "means of production" here is the
imperial system as a whole'. (Rex and Tomlinson, 1979: 13.)

The result was that, even in the early 1950s, prejudiced attitudes existed
towards coloured people in Britain. Hill (1965) was one of the first to demonstrate this fact empirically when he published the results of a survey undertaken
in 1951. Since that time, a series of similar studies has shown that prejudice
continues to exist and that, contrary to notions such as Bogardus's (1930) race
relations cycle, it may even be increasing (see e.g. Hill, 1965; Abrams, 1969;
Bagley, 1970; Richmond, 1973; Robinson, 1984a). In fact, the most recent
study of this type (Airey, 1984) has found that even when people are asked
directly whether they consider themselves to be prejudiced or not, over one-
third of people are willing to admit that they are. Furthermore, some 40 per
cent of respondents thought such prejudice would increase during the 1980s.
As one of the authors of the study commented (Airey, 1984: 125): 'In a society
apparently committed to racial equality and tolerance, we would expect the
proportion describing themselves as prejudiced to have declined. But we
wonder whether the opposite trend has not been establishing itself—with
expressed prejudice becoming more and more acceptable'.

One of the behavioural products of prejudice is discrimination, but it would
clearly be impracticable to either describe or quantify the extent of this
phenomenon in all aspects of British life. Instead, the remainder of this chapter
will concentrate solely upon the issue of housing, taking this as an example of
trends which are equally as apparent in the realms of employment, education,
access to services, etc. The argument which is presented is that ethnic prejudice
and discrimination add a specific competitive disadvantage to those general
handicaps which ethnic minorities already possess by virtue of their lower
incomes, larger average family sizes, and limited experience of British bureau-
cracy. In combination, these two sets of handicaps serve to disproportionately
disadvantage ethnic minorities in their struggle for access to scarce housing.
They have a more restricted choice of residence and residential location than do
their white counterparts. The net result is that members of ethnic minorities
form separate elements within the housing class structure and do not forge the
links with whites which might be expected on the basis of shared interests alone.
The behavioural concomitants of this are that ethnic minorities are spatially
and socially segregated and share few interactional links with members of white
society. Although the precise mechanisms by which this occurs are different,
similar conclusions apply to relations within the realms of education, employment, or access to the welfare services.

Immigrants and Housing in the UK

Why housing?

There are good precedents for selecting the interaction of immigrants with the
housing system as an example of more general trends, but it may well be

worthwhile rehearsing the arguments for this choice. Rex (1968: 214) states the thesis in its most extreme form: 'I suggest that the basic process underlying urban social interaction is competition for scarce and desired types of housing.' In Rex's (1973: 111) words 'any individual, but an immigrant in particular, "identifies", and is identified, much less with the place in which he works, than with the place where he lives', and in more detail (Rex and Tomlinson, 1979: 33) 'it is less likely that work relations will become the central focus around which other relationships, norms, and ideas cohere. Far more likely, as the focus of organization and identification which transcends institutional boundaries and truly communal solidarity, will be the neighbourhood and the defence of that neighbourhood against the world outside'. For Rex then, housing is the key arena not only because of its scarcity as a commodity but also because it forms the nucleus of social status and of patterns of meaningful social interaction and political activity.

The issue of housing as an indicator of social rank is, of course, a key one within urban geography, and there exists an extensive literature on social segregation brought about by the importance of address. In theory, the significance of address varies greatly according to the political system which controls a given society, such that address should imply little about an individual's rank in socialist cities but is a direct and conscious indicator of status in countries such as South Africa (Lemon, 1976)—much as it was in the pre-industrial city (Sjoberg, 1960). In practice however, address has status connotations in almost all societies, even those under socialist regimes (Bater, 1980; French and Hamilton, 1979; Musil, 1968). In the UK, where market forces are more dominant, address adopts the role of a 'public status symbol' allowing relatively instantaneous appraisal of any individual's position within the hierarchy.

However, the location of housing indicates not only social status and the amount of power wielded by an individual, but also the level of access to resources such as welfare facilities (Bell, 1977). This basic fact formed the touchstone of the wide range of area-based policies which were introduced in the late 1960s and early 1970s since each of these relied upon the assumption that individuals would use those facilities which were closest to their place of residence (see Eyles, 1979 for a review).

Lastly, a significant literature points to the relationship which exists between an absence of spatial distance (i.e. similarity of residential location) and the presence of social interaction. Such interaction ranges from a variety of forms of positive social interaction—for example intermarriage (Kennedy, 1943; Koller, 1948) and friendship patterns (Festinger, Schachter and Back, 1950)— through to extreme negative interaction such as crime (Smith, 1981). A clear statement of this relationship between spatial and social distance can be found in Peach's (1975) book *Urban Social Segregation* whilst Beshers (1962) argues the existence of a parallel relationship between spatial distance and similarities in values and morals.

There are, therefore, good *a priori* reasons for selecting housing as the issue which is likely to illustrate the general operation of those mechanisms which are responsible for allocating scarce resources to immigrants. Beyond this, it is also possible to argue that not only is the housing system representative of other allocative mechanisms, but it is also responsible for allocating *the* single most important scarce resource.

Empirical evidence of ethnic disadvantage in the housing system

A complete discussion of the structural context into which coloured immigration took place cannot assume that the ethnic prejudice described above necessarily translates itself into prejudiced behaviour by white people. As Allport (1954) notes, prejudice produces a range of behavioural patterns which stretch from anti-locution at one extreme to extermination at the other. Between these extremes lies discrimination, which Allport sees as the first active response to prejudice in the sense that the individual transfers the cost of his beliefs onto the group against which he is prejudiced, rather than bearing it himself. It is important, therefore, to demonstrate that prejudice really does generate discriminatory structural constraints within the housing system and that these constraints operate to disadvantage coloured migrants more than other deprived groups. Needless to say, this is a task which many have attempted, and there now exists a voluminous literature on the topic which can only briefly be summarized here.

Daniel (1968) was perhaps the first British researcher to study thoroughly the extent and form of discrimination. His report was based on three methods of assessment: interviews with coloured immigrants; discussions with institutions, companies, and local authorities who were in a position to discriminate in the provision of housing, employment, and services; and situational actor tests with companies, bureaux, and individuals. Daniel's conclusions with regard to housing were subdivided into those relating to private letting, home ownership, and council renting. His investigation of the rental market began with a survey of private landlords and his findings were as follows:

Ignoring both those flats that are not advertised and those advertisements that exclude him by stating 'no coloureds', he can expect discrimination in about two-thirds of his applications. Normally this will mean that the accommodation will be refused him altogether. But even on some of the occasions when he is not refused he can expect to be charged a higher rent . . . He can expect discrimination whether he is a bus conductor or a hospital registrar . . . Occasionally he will be met by brutal insults and, not infrequently, he will meet hostility and unpleasantness. (Daniel, 1968: 156.)

Daniel also discovered the 'reasons' behind such behaviour. Private landlords feared overcrowding and the deterioration of their property; they considered blacks to be noisy or dirty; they were conscious that they should not upset their neighbours; and they believed that the entry of black tenants into a neighbourhood would signal a decline in its status.

Letting agencies had to reflect these same views since they simply acted on behalf of the landlord and were not an independent force. Many of their properties came to them with specific instructions that they were not to be let to coloured tenants, and even those landlords who were less open about their prejudice *expected* not to be sent black applicants. When the agents did send coloured applicants they were berated for doing so, and were occasionally approached by residents who lived near the accommodation and who tried to persuade them not to do so again. Because of these reasons, and the low success rate for the placement of coloured applicants, many agencies deliberately dissuaded blacks when they enquired about tenancies, although they did not like doing so. Others directed blacks only to areas where they knew that coloured renters had already gained entry. Daniel summarized the position by noting that 'massive discrimination' existed in private letting based solely upon colour.

The study produced similar conclusions for the owner-occupied sector. Coloured applicants were treated differently by estate agents in well over 50 per cent of cases, usually either by being told that a mortgage would not be forthcoming or by being shown a limited number of properties. Again, the agents pleaded that they only acted as middlemen and reflected the wishes of their clients who shared similar fears to those of the private landlord. Daniel thought, however, that in some cases, agents did enjoy a degree of freedom and that this was not always used to the benefit of coloured applicants, who were often shown only older houses in the declining parts of towns. The estate agents themselves rationalized this behaviour by arguing that there seemed little point showing blacks more expensive properties when the financial institutions would not grant mortgages to coloured applicants. For their part, the building societies retorted that they treated each application on its merits, although it was clear that the financial yardsticks which they used *did* disadvantage coloureds, and that, in addition they regarded such people as unreliable. Daniel also used the evidence of demand for local authority mortgages as a sign that conventional forms of finance were not open to blacks, although he also pointed out that even this source was not free of discrimination since some local authorities demanded a five year residence period before they would consider an application.

Finally Daniel considered the public sector of housing where he found no overt attempts to debar black applicants. Nevertheless, he did argue that discrimination existed, and he cited, in support, cases of coloured immigrants being allocated patched housing (a point also raised by Burney, 1967) rather than purpose-built properties, being housed discreetly so as not to create any backlash from whites still on the waiting list, and having the demolition of their own houses delayed so as to postpone the need to provide council accommodation. Like the Cullingworth Committee (1969), Daniel also suggested that the application procedure, involving residence qualifications, the accumulation of 'points' and the subjective opinion of housing visitors, could work to the disadvantage of immigrants. He additionally noted at least one case where policies had been drawn up specifically to exclude coloured immigrants but had then been disguised as a more general and impartial policy.

On the basis of the survey evidence which Daniel (1968: 209) had accumulated he concluded that 'in the sectors we studied—different aspects of employment, housing, and the provision of services—there is racial discrimination varying in extent from the massive to the substantial'. More specifically, he noted that while 'some of the trends in employment are towards improvement in the position of coloured people, we found no indication of comparable trends in housing, where discrimination appears to be even more general and more massive than it is in employment' (Daniel, 1968: 219). As for the causes of discrimination, he argued (1968: 209) that 'the experiences of white immigrants, such as Hungarians and Cypriots, compared to black or brown immigrants, such as West Indians and Asians, leave no doubt that the major component in the discrimination is colour.'

As a result of the intervening legislation concerning race relations in the UK, Political and Economic Planning (PEP) commissioned a second survey in 1974 to gauge the extent to which overt racial discrimination had declined or increased in the seven years since its first report. Disappointingly the conclusions, presented in a series of reports which covered racial discrimination (Smith and McIntosh, 1974), public housing (Smith and Whalley, 1975), employment (Smith, 1974), and racial disadvantage (Smith, 1976; Smith, 1977) were very similar to Daniel's, although the 1974 data indicated a decline in overt manifestations of racial discrimination, if not actually in discrimination itself. Smith and McIntosh (1974: 35) thus note that 'the results of our tests demonstrate that there is still substantial discrimination against members of the minority groups when seeking jobs and housing. They also show that this discrimination is more related to race than to foreign origin . . .'. Encouragingly, they were able to point to a decline in discrimination within the field of housing. In particular, a significant decrease had taken place in the context of private renting. Whereas in 1967, 62 per cent of coloured applicants were discriminated against by landlords, this had fallen to 27 per cent in 1974. Smith (1977: 290), however, warned against misinterpretation of these data because 'it has to be remembered that the great majority of Asians, and over half of West Indians, are living in owner-occupied accommodation and will therefore have had no recent experience of seeking rented accommodation. Further, when they did look for accommodation to rent, they may have done so through personal contacts (probably of their own group) who they knew would not discriminate against them'. The 1974 survey also revealed a decline in discrimination by estate agents, although here again about one in five coloured applicants were still being treated differently from matched white applicants. The former were either offered an inferior range of housing to their white counterparts (in 12 per cent of cases), or a similar number of houses but ones which were located in different areas (in 5 per cent of cases). As for public housing, Smith and Whalley (1975) were careful to draw the distinction between racial discrimination and racial disadvantage, and their own explanation of the housing position of blacks was based more upon the latter rather than the former. They did discover, however, that residence qualifications were still employed by local authorities, and that in some cases, priority for housing was allocated on the basis of residence or waiting times. They also found evidence of actual racial bias in allocation procedures and differential rating of families on racial criteria by some housing visitors. They summarized their overall views on the allocation of public housing in this way: 'although local authorities have not *consciously* implemented discriminatory policies, or policies that are discriminatory in effect, some discrimination has occurred' (Smith and Whalley, 1975: 79). In total then, the 1974 PEP study demonstrated that the 1968 Race Relations Act *had* had an impact upon levels of overt discrimination, particularly in the sphere of housing. However, Smith was quick to point out that this should not be allowed to encourage complacency since, despite the improvement, racial discrimination (and racial disadvantage) still continued on a substantial scale.

More recently, Brown (1984) has led a third Policy Studies Institute study of the position of blacks and Asians in Britain. He chose not to follow the techniques of the earlier studies and did not therefore undertake any situational actor tests which could directly quantify the extent and form of racial discrimination. Instead the 1982 survey relied upon self-reporting by respondents, a method which Brown argues is likely to understate the real level of discrimination. He discovered that over two-thirds of West Indians thought that white landlords discriminated on racial grounds. One-third felt that council housing departments treated West Indians in a different manner from whites, and one-fifth of West Indian respondents thought that estate agents and banks were biased against them. Similar questions asked of Asian interviewees produced a more limited perception of discrimination with 34 per cent, 22 per cent, and *circa* 6 per cent considering that discrimination existed in those areas mentioned above. Perhaps more important than these general beliefs was the issue of whether respondents had themselves experienced racial discrimination when applying for privately-rented accommodation. 49 per cent of West Indians and 39 per cent of Asians claimed to have experienced discrimination. Returning to a more general level, Brown also found that over 40 per cent of Asian and West Indian respondents felt that discrimination in Britain was worse than it had been five years previously. In conclusion, Brown (1984: 318) noted that 'it is clear that racialism and direct racial discrimination continue to have a powerful impact on the lives of black people'. More particularly, 'irrespective of the differing needs of white, Asian, and West Indian households, there is a considerable gap between the quality of black and white tenants' properties. . . . Direct and indirect discrimination has played a part in this process, and continues to do so . . .'.

The evidence and conclusions of national surveys such as those of Daniel, Smith, and Brown on discrimination in housing have also received a good deal of support from a host of more detailed and specific studies, only some of which can be mentioned here. In the private sector, the work of Collard (1972; 1973) and Hatch (1971; 1973) has indicated the importance of exclusion by white estate agents who fear that coloured occupancy will result in blight. Cater (1981) has underlined this by demonstrating how separate housing sub-markets have developed in Bradford; Asian transactions being controlled by Asian estate agents, white transactions by white agents. Fenton (1973; 1977) has linked factors such as these with the higher search costs which Asians are forced to bear when looking for new property. Frequently these higher costs outweigh the degree of dissatisfaction with the existing property, and as a consequence the family remains in the inner city. Those who do pursue the matter eventually become willing to pay a premium for property in order to overcome racial discrimination. Fenton's evidence for Manchester suggests that this premium might be as much as 5 per cent of the purchase price. Even when a coloured family has been able to find a suitable property and persuade the vendor to sell, the problems are by no means over. Stevens *et al.* (1981) have shown that building societies, in Leeds at least, prefer to lend to white clients, whilst both Karn (1978) and Duncan (1977) have demonstrated the relative absence of society lending in inner city areas of ethnic settlement. Karn (1978) goes on to argue that other sources of finance such as banks or local authority mortgages are insufficient to compensate for the lack of building society involvement in these areas.

Discrimination and disadvantage are not, however, restricted to the private sector, Burney (1967) being one of the first to indicate that they might also occur in the public sector. In particular she noted two sources of bias in the allocation of council housing, notably the housing visitor's assessment of candidate suitability, and the tendency to give successful coloured applicants mainly older and inferior property. The latter point is one which has continually recurred in the literature over the last decade or so. The Runnymede Trust (1975) showed this to be the case in Greater London, where few families from the minority groups live in the outlying cottage estates, and most reside in higher density, centrally-situated flats (see also Flett, 1979, on Manchester). The same point was also made by Parker and Dugmore (1977), and by the GLC (1976) itself. The latter survey revealed that 44 per cent of publicly-housed blacks lived in older flats (compared with 15 per cent of whites), and only 24 per cent lived in houses or post-1964 flats (cf. 55 per cent of whites). Rex and Tomlinson (1979) noted a similar phenomenon in Birmingham during the 1970s. Karn's comprehensive survey of the allocation system in the same city indicated that little had changed between the early studies and the position at the beginning of the 1980s. Karn (1981) drew seven conclusions from her research: that despite the relaxation of residence qualifications, there are still significant constraints to access by Asians; that Asian applicants need more points than an equivalent white applicant before being allocated a house; that coloured applicants receive older housing; that the partiality of the transfer system exacerbates unequal initial allocations; that Asians list only a small number of estates on their application forms, possibly because of inadequate advice; and that coloured tenants are considerably less happy with their properties than are white tenants. Skellington's (1981) study of Bedford and Simpson's (1981) of Nottingham underline Karn's findings, whilst Peach and Shah's (1980) analysis of GLC data suggests the likely spatial impact of such discrimination and disadvantage, whilst the Commission for Racial Equality (1981) has pointed to the rise in violent attacks on members of ethnic minorities who live on council estates. More recently, the CRE's (1984) investigation of the allocation procedures of Hackney's council housing department have left observers in no doubt about the persistence of ethnic discrimination and disadvantage in the public housing sector and the scale and importance of this phenomenon. Notwithstanding criticisms of the methodology of the study on which the CRE based its recommendations (Harrison, 1984), it may well be that this investigation will prove to be a watershed and authorities may become more aware of intentional and unintentional discrimination in their allocation procedures. Even if this does prove to be the case, it is still likely to be some years before the effects of any reforms are fully felt.

The structural product of ethnic discrimination and disadvantage

It remains to consider what the ramifications of discrimination and disadvantage really are for the overall positioning of ethnic minorities in the housing market. This is particularly important given the contention that housing forms a key factor in determining individual and group status, access to services, and opportunities for social interaction. Rex and Moore's (1967) housing class model is perhaps the best known and most stimulating attempt to locate ethnic minorities in the housing hierarchy and to consider the social implications of this.

According to Rex (1968) the idea of housing classes grew out of a desire to extend the ecological approach of the Chicago school. He began by making several assumptions about the housing market and housing demand: 'The model we have posited assumes the existence of a socialist movement in relation to housing amongst the native working classes, an inability to exercise political power on their own behalf by disadvantaged groups, and an aspiration to relatively detached family life in suburban conditions amongst all groups' (Rex, 1968: 216). Like the ecological school then, Rex and Moore considered that the character of residential mobility within the city was predominantly centrifugal in search of idyllic detached property in a rural setting (Moore, 1969). However like most other valued commodities, the supply of suburban housing, either in the form of peripheral council, or owner-occupied, properties, is insufficient. Rex consequently suggested that one of the basic features underlying much of the social interaction which takes place in the city would be competition for, and conflict over, access to prized suburban housing. In this competition it was assumed that different individuals or groups would have different chances of success because of the unequal availability of economic or political power. The most favoured or powerful groups would thus ensure their own access to suburban housing and attempt to prevent the access of others. Those without the necessary strength to overcome such exclusion would be forced to settle for the next most desired property, and so on until the least powerful group, who would have no choice in their housing, would be constrained to occupy those types of houses or those areas which had been discarded by all other groups in the hierarchy. Relying on the Weberian notion of class, Rex and Moore then proceeded to classify individuals by the quality of their present housing, which was seen as a clear indicator of their access to power and their current position in the social hierarchy. Rex went further than this though, and predicted that under certain circumstances, shared membership of a similar housing class would produce a class-conscious grouping with concomitant patterns of intra-group relationships and interaction. In short, a class-in-itself would become a class-for-itself, and could therefore be mobilized to further its advantage or to defend its existing position (Rex, 1973). Rex was thus arguing that housing and access to housing generated a class system which did not necesarily parallel that created by economic production (Rex, 1977). The analysis of a person's housing situation could thus provide clues to his social status, his access to power, and his likely patterns of social interaction which an isolated analysis of economic class membership alone might overlook.

In the 1967 study, Rex identified a hierarchy of six housing classes each of which had its own entry requirements. The classes and entry qualifications were as follows:

(i) outright owner occupation: is highly prized but presupposes large cash savings or fulfilment of mortgage repayments after a lengthy working life.

(ii) owner occupation by borrowing from a building society or bank: a popular strategy but one which requires the availability of a cash deposit, a regular and substantial income to repay the mortgage instalments, and a willingness to maintain the property (to safeguard the society's loan) and not to sub-let parts of it to lodgers.

(iii) public housing: in theory allocated according to need to those households unable to fulfil the entry requirements to (i) or (ii) but, in practice, biased towards applicants who conform to the housing norms of the bourgeoisie.

(iv) renting a complete house from private owners: access extremely limited since tenancies are rarely advertised and often go directly to kin or friends of the landlord.

(v) ownership of a lodging house: an undesirable and illegitimate form of housing which society treats in a paradoxical and contradictory manner. On the one hand, it is seen to provide an essential safety net in the form of a stock of accommodation for those people whom society is presently unable, or unwilling, to house in a legitimate way. On the other hand, it is condemned by all as the product of naked capitalism and personal greed. As a consequence, regulatory bodies sometimes attempt to incorporate and legitimize lodging houses whilst at other times they try to outlaw them and persecute their owners. The entry qualification to this class is simply an ability to raise a cash deposit (often from relatives or loan companies).

(vi) tenancy of a room within a lodging house: also illegitimate and undesirable. The only entry qualification is an ability to pay a relatively high rent at fairly regular intervals. The absence of social or personal qualifications attracts those individuals who would not be made welcome elsewhere because of some socially defined non-conformity. Rex and Moore did also admit that some groups might *choose* to live in such conditions even when they need not do so.

Having proposed this model of housing classes, Rex and Moore were concerned to see how coloured immigrants might fit into this structure, and whether any housing class consciousness developed between those immigrants and indigenes who shared similar housing situations. To take each class in turn:

(i) owner occupation, either outright or through a mortgage. Rex and Moore (1967a: 37) allowed that 'so far as immigrants are concerned, access to the mutual aid system which operates in their own primary communities, together with their willingness to work overtime, makes it possible for some of them at least to obtain a deposit'. But they argue that vulnerability to redundancy, obligations to kin, and style of life all mean that 'the number of immigrants who succeed in entering the class of owner occupier is small'.

(ii) tenancy of a council house. 'Most immigrants are excluded on grounds of residence alone, but those who do qualify on these grounds are likely to be admitted only to the inferior council houses. The upper sub-class of council tenants in council built flats and houses will include very few immigrants.' (Rex and Moore, 1967a: 37.)

(iii) tenants of whole private houses. 'The private landlord has the power to discriminate according to his own whims. Since nearly all landlords who let whole houses are English, the overwhelming likelihood is that these tenancies will not be given to coloured immigrants.' (Rex and Moore, 1967a: 38.)

(iv) lodging house proprietors. 'It is easier for an immigrant to obtain these qualifications than to acquire those which are essential for any of the three classes mentioned above.' (Rex and Moore, 1967a: 38.) Reference was clearly being made here to the availability of loan funds from relatives (Asians) or through 'boxes' of rotating credit (West Indians).

(v) lodging house tenants. Since this class is constituted specifically of people who have had their access to other forms of housing barred by their inability to meet entry qualifications, it is likely that immigrants will be considerably over-represented. As Rex and Moore (1967a: 38) put it, the class will contain 'immigrants, many people with irregular forms of family life, and social deviants'.

   In summary, because of their recent arrival in the country, their position as a replacement labour force, and their poverty prior to immigration, immigrants are denied access, on the whole, to any housing other than that provided by the lodging house. In Rex's words (1973: 8): 'For, after the various sorts of selection essential to the allocation of "desirable housing" have occurred, there are bound to remain some who simply cannot qualify and are, therefore, in the desperate position of being willing to settle for precisely the sort of accommodation which the lodging house landlord has to offer'. In terms of spatial segregation, because of the way in which cities have developed organically, immigrants become concentrated in the Zone in Transition where houses are blighted prior to redevelopment and are therefore available on short leases at a low price. Even those migrants who have succeeded in gaining access to council housing are likely only to be allocated patched properties which are frequently also located in the Transitional Zone. The spatial distribution of the housing stock and the operation of the housing market thus confine immigrant populations to involuntary residential segregation. The need to conserve funds may also translate residential segregation into workplace segregation, with immigrants unable to afford the lengthy journey to the surburbs (see McCormick, 1981, on this).

Having identified the outlines of the housing class system and seen how, in objective terms, the immigrant population fitted into this hierarchy, it remained for Rex and Moore to consider how the actors involved in the system subjectively viewed their own position and that of black residents. Furthermore, it was important to ascertain whether such perceptions induced a behavioural response. Rex and Moore chose to investigate these issues in the context of the Transitional Zone of Birmingham, namely in Sparkbrook.

They discovered that ethnic solidarity, stereotyping, and the cross-cutting influence of associations all tend to mute class conflict at a local level to the point where life is characterized by a fragile peace rather than overt conflict. But, as Rex (1968: 14) warned, 'a sudden incident could polarize the situation, and, in any case, the overall effect of the urban encounter of the various ethnic groups has been to exacerbate whatever tendencies to conflict were present between them already'. Out of all this, immigrants appeared as a pariah group and an underclass. They became pariahs because they undertook a task which, whilst it was essential, was regarded as socially undesirable—the provision of cheap housing for both their fellow immigrants and the social deviants who fell outside the state system. Their subdivision and subletting of Transitional Zone property marked them off, in the eyes of the indigenous population, as destroyers of good housing and benefitters from other people's misfortune. They were harried and pressured by the state, but never to the point where they ceased to provide accommodation altogether and forced the authorities into having to make alternative provision for their tenants. In short, the competitive struggle for scarce housing therefore concretizes, legitimizes, and personalizes the vague notions of prejudice against black people which existed in Britain as a legacy of colonialism. It gives reason to prejudice. The reaction of the black population to what they see as rejection by white society takes the form of refusing to identify with white working-class culture, community, or politics and instead creating a parallel series of associations and lifestyles unique to the black underclass.

Beyond the issue of housing, Rex and Moore's thesis also provides an elegantly simple description of the way in which the variety of individual constraints and passive constraints build into a solid edifice of institutionalized discrimination tacitly supported by the majority of the white population. It also demonstrates how such institutional constraints maintain a barrier between white and black populations, and how the latter is forced to accept a *de facto*, if not *de jure*, inferiority.

Despite the clarity and appeal of the thesis, and the way in which it argues from the specifics of housing to the broader issue of overall social status, Rex and Moore's ideas have not gone unchallenged at either the conceptual or empirical level. A detailed review of these criticisms and the responses of Rex and his co-authors is available elsewhere (Robinson, 1982a), and need not be repeated here since the notion of housing classes is used simply to illustrate how the struggle for a scarce resource such as housing crystallizes attitudes and has a direct bearing on the overall view which society holds of a particular migrant group.

Conclusion

This chapter has looked at the social conditions which faced West Indian and Asian migrants on their arrival in Britain. More particularly it has studied the views that its constituent people had towards coloured migrants and discussed how these views are frequently translated into action and become codified into barriers designed to prevent the social mobility of black people. In doing so, it has concentrated upon one particular resource and indicator of social status, in the hope that this specific and detailed example will provide an insight into more general tendencies which exist in all spheres of societal activity. Housing was selected as an appropriate subject for such a study and reasons were given for this choice. A general discussion then followed of the way in which specific constraints within the housing market individually serve to block access to that housing which the white majority deems to be desirable. The ways in which these constraints combine and interact to produce a housing hierarchy which consistently and systematically bars access to its upper reaches to anyone other than 'deserving' and conforming cases was then outlined with the help of Rex and Moore's housing class notion. This notion had the added advantage that it considers the implications of competition for housing on other spheres of inter-ethnic contact and more generally upon the way in which coloured immigrants are seen by white society. The study of the allocation of scarce housing consequently provides a generalizable view of how white British society behaves towards coloured immigrants.

CHAPTER 5

Cultural Forces

The previous chapter was concerned solely with the context of those structural constraints within which coloured people in Britain must operate. All of these forces potentially constrain their behaviour and their access to scarce resources, as was demonstrated by the example of access to housing. The data clearly indicate that Asians are systematically excluded from 'good' housing and that they are allotted a role by white society as a marginal pariah group. However, such analyses are only one side of a complex argument. White society assumes, in an utterly ethnocentric manner, that all immigrant groups share its own aspirations, priorities, and value-system. Since housing is a valued status symbol in western society, access to it is restricted, vetted, and manipulated by appointed gate-keepers, and the system of allocation is rationalized by reference to stereotypes. Since detached, suburban property is the most highly prized, access to it is the most heavily controlled. In British society, attitudes towards defensible territory encourage the idealization of the detached property standing in its own grounds. Concepts of personal space (Lyman and Scott, 1967) lead to a positive evaluation of low density living and 'privacy', both of which require large properties where each individual may have his or her own territory or room. Status considerations require major investment in double glazed windows, patio doors, stone cladding, double garages, fitted kitchens, multiple bathrooms, central heating, barbecue areas, etc., etc. But in all these cases, society, or the individual within it, has made value-judgements about what is desirable and therefore what has to be prized and protected. Few stop to think about whether these same resources and characteristics figure as highly in other groups' lists of priorities or whether in fact they are protecting 'assets' which are deemed to be of secondary importance by their supposed competitors.

The purpose of this present chapter is therefore to look in detail at the sending societies from which most Asians have come to Britain and the attitudes and values which migrants are likely to have brought with them. Such an approach is valuable because it indicates the extent to which a migrant's orientation is likely to bring him into conflict with structural constraints, and therefore, whether such constraints remain only at the potential level or whether they are realized. It is important to note the caveat, however, that one should not be seduced by the superficially attractive option of regarding the structural forces operating within the receiving society as 'constraints' and the cultural values which the migrant brings with him as producing 'choices'. On closer analysis such a distinction clearly rests on rather arbitrary categorizations, for cultural forces which impel an individual to undertake a given course of action are obviously as much 'constraints' as are institutional racism or racial disadvantage. Conversely, no matter how restricted is one's structural niche there is always room for 'choice' between alternative courses of action although some may involve greater penalties than others. Also, it is unreal and unwise to regard 'choices' and 'constraints' as being two independent and mutually exclusive categories since the two are more often than not interacting and exist only in relation to each other.

A second caveat is also of some importance. Whilst an analysis of structural forces alone is without doubt an imperfect route to understanding contemporary race relations, research strategies which rely solely upon the actor's point of view are equally deficient. They place culture and individual choice on a false pedestal and ignore the very real restrictions which shape the life of each and every individual in a society. The Ballards (1977: 53) clearly recognized this pitfall when they wrote:

it is the external constraints of discrimination which set the limits within which South Asians . . . may operate . . . the particular behaviour of different groups can only be finally explained in terms of the culturally determined choices made within these limits, as well as the various ethnic strategies used to counteract, circumvent, or overthrow these constraints.

It is to these issues that the remainder of this chapter is devoted.

Social and Economic Conditions in the Sub-continent

It is impossible, in such a small space, to describe fully even the major social and economic characteristics of societies with such long and complex histories as those of India and Pakistan. Nevertheless, it *is* possible to describe the major societal elements which have had an impact on stimulating overseas migration and those which are most likely to influence the behaviour of migrants once abroad.

Some social considerations

Perhaps one of the most important elements in the structuring of Gujurati and Punjabi societies is the combined factor of the joint family and respect for elders. The extended family not only ensures a high degree of social control over its younger members but it also emphasizes a common family identity and therefore a common status. In the sub-continent, the joint family also has economic implications in that members are often participants in the same economic enterprise, notably the smallholding: each individual is expected to contribute similar inputs in the form of labour or, far less frequently, capital. Any benefits which accrue from joint efforts are held by the head of household on behalf of the family. As a result, both in economic and social life, the family head retains total authority, and demands the loyalty of each and every household member. Consequently, life is structured for each family member so that roles are clearly defined and the individual subordinates his own will to that of the group. Clearly a system such as this cannot operate unless there is an in-built, or socially sanctioned, respect for authority.

The second feature worthy of note is a concern with status or respectability, and a concomitant drive towards achievement. This may be manifested at an individual level but it is more likely to be active at the household or group level; Pakistanis for instance think very much in terms of the *biraderi*; a group of individuals who share common views on morality and religion, and frequently act as a unit. *Izzet* (status or social worth) is regarded very much as a fragile commodity which may be destroyed by a relatively simple act, and Punjabis are consequently very aware of the possible implications of their actions on other members of their joint household or *biraderi*. This is particularly so for many men who still have unmarried sisters, for their wayward actions impinge directly upon their sisters' marriageability. Indeed some commentators have argued that it is the need to know the potential spouse and her family very well (in order to ascertain their relative status) that necessitates *biraderi* endogamy amongst Muslims. Without such prior knowledge it would be much easier to lower one's own family *izzet* by unwittingly marrying into a family of low status or dubious honour. Thompson (1970) comments upon similar fears amongst Punjabi Sikhs. Clearly many of the actions of Punjabis are constrained by a knowledge that they must not, under any circumstances, jeopardize the status of their joint household and *biraderi*. However one can go further than this and argue that the concern with *izzet* acts as a positive force to encourage achievement and aspiration. One route towards enhanced status is the accumulation of wealth and this is frequently in the form of land, because the ownership of land not only has an economic value but a social value in its own right. The size of a joint family's landholding is, thus, often taken as a direct statement about their social standing. Pettigrew (1972) confirms the importance of land ownership because she comments about both the desire to be thought of as a member of the *zamindar* (landowning) castes rather than the *kama* (labouring) castes, and the social disgrace of not owning land. Ballard (1978) concurs, but she also mentions an alternative strategy for the enhancement of status through contracting prestigious marriage alliances. However, in order to attract a groom from a family of higher status, the bride's parents need to be able to offer a sizeable dowry: again this necessitates the accumulation of wealth, and if the approaches are rebuffed, this can itself harm the *izzet* of the bride's family. Finally, Jeffery (1973) described how *izzet* may be enhanced through the pursuance of particular lifestyles: she categorizes these as 'Ashrafization', and 'Westernization' in the case of Muslims, whilst Srinivas (1967) terms them 'Sanskritization', and 'Westernization' in the context of Hindus. 'Ashrafization' is characterized by a number of features: conformity to the Islamic model and therefore reference to the Koran for guidance on lifestyles; male occupational mobility into either professional or clerical positions; conformity on the part of women who should be modest, house-orientated, and remain uneducated in the formal sense; and an emphasis upon the joint household. The social style of the household should also be centred around devotion to prayers and fasting, but should remain generous and welcoming to guests. In contrast, Jeffery characterizes Pakistani 'Westernization' in terms of the presence of servants, the partial liberation of women, and sumptuous family parties at which alcohol is served. Women should be educated, and should dispense with wearing the *burqa*, whilst they should also be allowed to mix freely with members of the opposite sex at clubs. Clearly both life-styles require considerable resources, and this again is a force encouraging the Muslim towards the accumulation of wealth and the drive towards achievement.

Srinivas (1967) demonstrates that the same may also be said of Hindus, although the life-styles to which they aspire differ significantly from those of the Muslim. 'Sanskritization' is described by Srinivas (1967: 28) as follows: 'One of its functions was to bridge the gap between secular and ritual rank. When a caste or section of a caste achieved secular power it usually also tried to acquire the traditional symbols of high status, namely the customs, ritual, ideas, beliefs, and life-style of the locally highest castes.' In other words, for a caste to gain significant social mobility in the ritual sense, it must first register social mobility in the secular sense. The latter is often measured, as we have mentioned above, in terms of land ownership. As Srinivas (1967: 12) puts it: 'landownership confers not only power but prestige, so much so that individuals who have made good in any walk of life tend to invest in land. If ownership is not always an indispensable passport to high rank, it certainly facilitates upward mobility.' In short then, only when a caste has gained dominance of local landownership can it hope to have its claim to higher status in the caste hierarchy treated seriously. An alternative route to social mobility for a Hindu is through 'Westernization'. Srinivas does not detail the specific characteristics of Hindu 'Westernization' in the way that Jeffery does for Muslims. He prefers to think in terms of themes or orientations, and he considers the dominant theme of Hindu 'Westernization' to be secularization of all aspects of life. Whilst much of this rests upon permissive legislation and institutional change, Srinivas does give some insight into the material changes which have necessarily accompanied the new ethos. He cites in particular the decline in ritual during the process of food consumption, and incidentally notes how simply eating from leaves whilst seated on the floor has been replaced by such paraphernalia as dining tables and chairs, plates and cutlery. He notes also the emphasis upon large marital dowries, upon the education of males and females, and the attraction of new clerical careers which necessitate qualifications and a secularized daily routine. Both at an individual and societal (village/regional) level each of these changes requires investment in facilities and therefore underlines the importance of saving and aspiring.

In summary, then, for both the Hindu and Muslim, higher status (or *izzet*) can only be achieved through the possession of material wealth either for itself or as a means to an end.

Environmental conditions

Two major studies by anthropologists tell us a good deal about social and environmental conditions in the sending areas. However, one must be alive to the fact that these studies were undertaken some time ago and that Indian and Pakistani societies are currently undergoing considerable change. Some sources of this change will be discussed later, but in the absence of more recent work the general account presented below must, perforce, rely upon the insights gained by Dahya (1973; 1974) in the early and mid-1960s, and Jeffery (1972; 1973; 1976) in the later 1960s. Their findings were, however, validated by the author's field visits to the major sending areas in India and Pakistan during 1982.

One of the issues which is likely to have a bearing both on how an immigrant group is received, and how quickly it is able to gain social mobility is the similarity of its educational standards to those of the host society. Jeffery has shown how access to education in Pakistan is variable: males always receive priority over females; the rich are better able to educate their children than the poor; and the urban dweller is better provided for than the rural. Marsh (1967) indicates parallel discrepancies in India but also shows that these are overlaid by religious differences such that the Parsees are the most literate whilst the Muslims are the least.

The formal data on employment and social structure are also very patchy but the few details which are available do give a clear indication of conditions. According to Jeffery, approximately 80 per cent of the Pakistani population are resident in the villages, and even of the 20 per cent who are not, many are only recent rural–urban migrants. One should not underestimate the social importance of the village, for, as Khan (1977: 62) notes, 'the village forms a moral arena in which reputations are assessed and re-assessed, and potential deviants pulled back into line'. Rose (1969) considers another facet of village life when he cites information which suggests that well over half the rural population is underemployed. Those who are fortunate enough to find regular work in rural areas are naturally concentrated chiefly in the primary sector, although others are craftsmen or professional/service workers. John (1969) describes the social structure which develops from this employment base in a Punjabi village, and notes how there is a graded hierarchy from the rich landlord, through the owner-cultivator, tenant, and the artisan, to the landless labourer. Srinivas (1967) estimates that the latter form approximately one quarter of the rural workforce, although this is now likely to be an underestimate. For those who are in full-time employment, remuneration varies a good deal between countries and between regions, and, again, some examples will have to suffice as an approximate guide. During 1982, field visits to thirteen villages located in the Indian Punjab, in Mirpur, and in the Kheda district of Charottar, produced a range of declared incomes between £150 per year for a *kami* to £1250 per year for a wealthy farmer near Jullundur City.

As for the issue which concerns us most here, that of housing, both Jeffery and Dahya give us detailed and authoritative accounts. Here, as much as in any other area mentioned, there are major discrepancies between the conditions which prevail in rural and urban areas. These conditions are symbolized by the characteristic *kutcha* house of the village and the *pukka* house of the town. The *kutcha* (or *kachcha*) house is a single-storey building constructed of sunbaked mud bricks. It is laid out around an open courtyard which is frequently used for the preparation of meals. All of the house is flat-roofed and during the hot season this area doubles as sleeping quarters for the male members of the household. Within the house, both of the two rooms are multi-purpose and often contain no furniture whatsoever. Few of the houses have piped water or mains sanitation, and it is normal practice for a household not to possess a water closet or bathroom. Despite winter temperatures in Pakistan which fall well below freezing, there is rarely any form of heating other than an occasional paraffin stove. Electricity is becoming more common for lighting but hurricane lamps still predominate. As for residential densities, Dahya (1974: 108) comments, 'overcrowding is a permanent phenomenon and forms part of the villagers' experience. Gregariousness in everyday life is valued because, among other things, it helps to keep a check on a person's movements and is a means of social control'. The *pukka*, or *pakka*, house is dissimilar in a number of ways and is characteristic of urban areas and more affluent villages. It is usually of brick construction and is therefore more substantial and permanent. Many are three storeys high although the upper two often remain unused but fully furnished. Each house will have water and electricity supplies laid on, and the roof houses a latrine. In addition, the house will exhibit greater decoration in the form of trellised balconies, loggias, and multi-coloured glass windows. However it is essential to note that both the *kutcha* and *pukka* houses are owner-occupied: only those too poor to build even a *kutcha* house are forced into renting.

Responses to Social and Economic Conditions

Although the causes of Indian and Pakistani emigration have already been briefly discussed in chapter 3, it is valuable to consider the topic again, not in terms of a deterministic push–pull model, but viewing migration as a response to blocked social mobility. Srinivas (1967) argues that social mobility was a significant characteristic of pre-British India and that that society was more 'open' than an initial analysis of the caste system might suggest. Srinivas writes extensively on the sources of such mobility but selects the labour market and the political arena as two key areas of opportunity. However he also argues that the scale and form of mobility changed greatly with the coming of the British Empire, and the societal transformation which this engendered. British rule froze the political system and closed this as an avenue of advancement. Demographic changes produced a burgeoning population which affected the availability, security, and remuneration of employment, and restricted mobility through the labour market. Therefore, whilst the quality of life for the lower caste worker might arguably have improved under British rule, his prospects for advancement most certainly did not.

Set against declining opportunities were the new aspirations which the British introduced, firstly to the local élite but latterly to the majority of the population. According to Srinivas, Westernization replaced Sanskritization, and individual achievement by the joint household replaced group ascription. However the difficulty with these new possibilities was that they were available only to the wealthiest since one needed education to benefit from new governmental opportunities, capital to benefit from new agricultural systems, and mercantile expertise to gain from the rapid expansion of overseas trade. In short, Srinivas's thesis is that British rule raised expectations for the majority but provided opportunities only for the few.

It is not too fanciful to see migration as the product of this paradox. Joint households seeking Westernization would initially despatch a son to the city to earn a living which could supplement income from the landholding. However, for several reasons this strategy proved inadequate for many households and indeed for entire regions: Muslims and Christians in India suffered from religious discrimination in job allocations; the Mirpuris received chiefly cash compensation after the construction of the Mangla dam and were therefore forced into the competitive market for increasingly scarce agricultural land; the Punjabi Sikhs were suffering enormous pressure from very high population densities, the absence of primogeniture and the fragmentation of holdings; the Gujurati Hindus resented the absence of sufficient urban or mercantile opportunities, for they were more educated than most groups (Brooks and Singh, 1979a); and the Campbellpuris hailed from an environmentally deprived area with a hostile climate (Rose, 1969). When one remembers that each of these groups had gained a good deal of contact with British influence either in the army, merchant navy, or as a trading element in other overseas colonies, it is not difficult to see how British colonial rule created new aspirations within these groups which India, and Indian society, were unable to fulfil, at least in the short term. It was not unnatural, therefore, that members of each of these regions should contemplate the strategy of sending sons overseas to benefit the joint household from the higher wage levels and better educational opportunities which existed in developed countries. Although the initial cost of emigration would take some time to recoup, the son would soon send sufficient remittances to achieve the desired status and hence allow his return to the sub-continent. The strategy was frequently predicated upon knowledge of economic opportunities gained by adventurers (see chapter 3) and those who were forced to work overseas by the very nature of their jobs.

Why Britain should have been chosen in preference to other countries remains an under-researched question, but undoubtedly the knowledge of economic opportunities existed and the labour shortages in certain industries must have seemed attractive. Additional factors would include the relative cost of living and wage levels, aspects of which have already been touched upon.

However, one should perhaps not overstress the positive factors of migration to Britain for negative factors did indeed exist. Smith (1976:20) notes that:

the disadvantage of coming here (compared with moving to the town within the home country) is that it means becoming a member of a small minority within an extraordinarily alien culture. A move from the countryside to the town within the home country must make great demands on people's powers of adaption; but the contrast between life in an Indian or Pakistani village and life in an industrial town in the North of England is so great in every way that the people who decide to make such a move must do so out of urgent need.

Secondly, Lawrence (1974) comments upon the fact that few Indians and Pakistanis actually wanted to come to Britain in the first place, but that they were forced to by their failure to gain entry to preferred countries like Canada and the United States.

The notion that international migration was an alternative economic strategy to that of internal rural–urban migration, and therefore that the former attracted those individuals and groups whose material satisfaction and prospects could not match up to the new standards generated by imperial rule, receives a good deal of support from surveys undertaken in Britain. Work by Dahya (1974), Rose (1969), Thompson (1974), Smith (1974), and Marsh (1967) shows that it was not the most destitute members of regional societies who migrated to Britain as a last resort, but those who wished to enhance or speed their social mobility through the accumulation of greater wealth or education. Many households were not devoid of status in the sending society but their aspirations encouraged them to seek more, either through Sanskritization, Ashrafization or Westernization.

Migration Intentions

If migration really was seen by South Asians merely as a means to improve the status of the joint household in India or Pakistan, then it is likely that settlement in Britain would be seen as a temporary expedient rather than as a desirable long-term state. Work by British researchers tends to show that this was the case, especially during the early phases of the migration. It has to be noted, however, that much of the research on this topic is either qualitative or relates to an unspecified 'coloured' sample.

Brooks (1969) produced one of the most widely quoted pieces on the intentions of migrants, based on a sample of London Transport employees which included West Indians, Indians, and Pakistanis. Data are not, however, broken down by ethnic group but the overall findings were that, on arrival, 44 per cent of individuals intended to stop in Britain less than five years, 37 per cent hoped to return eventually to their homeland, and only 4 per cent intended to settle permanently. Lawrence (1974) also found that less than 5 per cent of his Nottingham sample had originally envisaged permanent settlement in Britain, although Israel's (1964) Slough sample produced a result nearer to 12 per cent. Other authors have completed work which is less quantitative, but relates more specifically to the Asian population. John (1969: 19) thus writes of the Punjabi Sikh population: 'most of them planned, at first, to stay in Britain for only a few years—long enough to earn sufficient money to buy more land for their families back in Punjab'. Thompson (1970: 30) agrees: 'The first generation Punjabi immigrants in England have not, in general, come to make England their permanent home, but to fulfil certain economic goals and return to India'. The Ballards (1977: 27) also support this view: 'emigration has been regarded as a temporary move, an episode before a final return to the village'. Moreover, they suggest that this applies not only to Sikhs but to all emigration from the rural regions of the sub-continent. Dahya (1973: 245) tends to confirm this, for his work amongst Pakistani Muslims leads him to believe that 'the migrants consider themselves to be transients and all they demand of the host society is that they be allowed to work and earn for their families to whom they mean to return'. Jeffery's (1976: 62) work produced similar conclusions for the same group. And finally Lawrence (1974:27) concluded for a mixed sample of Indians and Pakistanis that 'there is no reason to suppose, therefore, that migration necessarily meant, or was thought of as, the first step on the way to a new and permanent life in Britain. Indeed, on the contrary, almost all of those interviewed intended to return home eventually.'

Notwithstanding the similar evidence which exists for the West Indian population in Britain (see Davison, 1966; Rex and Moore, 1967; Rose, 1969; Philpott, 1968) and the caution which is needed when interpreting responses to questions about sensitive issues such as settlement, it seems clear that those migrants who came direct to Britain from India or Pakistan initially viewed residence here as a temporary, economically-motivated expedient. In Richmond's (1968) terminology they were consequently not settlers but 'transilients' or transients.

Attitudes to British Society

Again it is difficult to summarize such a broad topic briefly, but several pieces of work provide an indication of how Indian and Pakistani migrants are likely to have viewed British society. A careful reading of Forster's (1924) *A Passage to India* is most illuminating here, since it highlights in a most sensitive but detailed way, those facets of British society and British character which were most on display during the colonial era. Moreover, it also highlights the power structure of colonial society and how this influenced relationships and attitudes between the indigenes and the British. Work such as Forster's provides important insights because, for many Indians and Pakistanis, the colonial period and its administrators were their only contact with, and experience of, the British. For many people then, 'British' became synonymous with a stereotype of the Turtons. Jeffery's (1976) work amongst the Muslim and Christian Pakistanis in Bristol tends to confirm this. She writes (1976:94) 'people in Pakistan have ideas about an undifferentiated ''British way of life'' . . . Colonial history and the behaviour of the British in India, the showing of British and American films . . . mean that people in Pakistan have a clear (if often erroneous) view of British culture'. She goes on to give a detailed account of the negative evaluations which many of her respondents hold about life in Britain. In particular, she notes criticisms of British morality, sexuality, life-styles, and family life, although it has to be said that her respondents *might* be expressing views which are somewhat more critical than the norm for all South Asians. Personal fieldwork in the sub-continent, along with Ghuman's (1980) work on the Bhattra Sikhs of Cardiff, suggests that this is *not* the case, although more research is required before any definitive conclusions can be drawn. Regardless of the exact strength of such negative evaluations, their true importance lies in the fact that most Indians and Pakistanis do not see British culture as a desirable or even viable alternative to their own. Faced with the choice, most opt to cherish and safeguard their own values and life-styles rather than adopt British equivalents.

The decision to retain the culture of the homeland is not, however, taken simply because of the unavailability of an attractive alternative. There are strong positive forces which lead South Asians to evaluate their own cultures very highly. Foremost amongst these is the migrant's belief that he will eventually return home to his village of origin (the 'myth of return'), and he is therefore encouraged to maintain strict conformity whilst abroad in order to ease his acceptance back into the community. Village life and its underpinning culture become idealized to the extent that the influence of the village is felt in all aspects of life in Britain, from the trivial to the crucial. Furthermore, despite the geographical separation of village and villager, the former does not lose its ability to inflict sanctions on the latter if this should prove necessary. Gossip forms the medium of communication and regulation, whilst the economic investment which most migrants have in their village of origin provides an added inducement.

Many South Asians consequently arrived in Britain with negative evaluations of that country's culture, and good social and economic reasons for retaining their own. However these predispositions could have been challenged and overturned if British society and its constituent peoples had acted in such a way towards the new arrivals that stereotypes were proved false to the extent that a re-evaluation was shown to be necessary. If such a re-evaluation had seriously challenged the attraction of the village, and therefore the 'myth of return', it seems likely that the behavioural consequences of the latter would have been only a temporary phase in the long-term process of adjustment. However, the literature makes it plain that the reception which British society gave South Asian immigrants was characterized by hostility and avoidance. Overt discrimination was widely perceived by Indians and Pakistanis and became a major source of disappointment. This would be unlikely to encourage deviation from adherence to traditional cultural standards and beliefs and, most probably, made life in Britain, and British people, appear even less attractive than they had seemed prior to migration.

Behavioural concomitants

The myth of return, the wish to avoid 'cultural contamination', and the maintenance of traditional values are all likely to have influenced the strategies adopted by the early South Asian migrants and therefore their patterns of behaviour. The behavioural responses to these cultural preferences are discussed in the following section.

A domain which is of central importance to this book is housing, and since it also represents the first problem which faces a new immigrant on arrival in a country it seems an appropriate starting point. Dahya (1972; 1973; 1974) has been the author who has most forcefully articulated the thesis that cultural preferences provide the key to an understanding of the housing situation of

South Asians in Britain. His arguments merit a detailed analysis, but they must be reconciled with the knowledge that cultural preferences operate only within the context of structural constraint outlined in the previous chapter. Dahya's (1974) arguments related initially to the quality of housing which South Asians would demand. He suggests that this would be determined by three considerations. Firstly, South Asians hoped to return to their homeland relatively quickly with their accrued savings. Anything other than minimal housing would divert resources away from this pre-eminent aspiration. Secondly, since South Asians associated landownership with status, renting would represent only a short-term expediency, especially when owner occupation was often cheaper. Thirdly, for a group who judged accommodation in Britain by standards prevailing in India or Pakistan, even the worst housing available in British cities compared favourably with the *kachcha* houses back in the village. In addition, Dahya contends that even the location of ethnic housing in inner cities can be seen as part of a culturally-determined strategy dominated by the need for frugality, rather than comfort. Access to kin, transport facilities, and employment drew the Asian population to the inner areas of Britain's cities in preference to the space, privacy, and status of the suburbs. For Dahya's Pakistanis, then, the inner-city terrace represented an ideal solution to their needs since they did not seek immediate gratification in Britain but rather a deferment of this until after their return home (see also Avison, 1965). The Ballards have since extended this argument to Asian groups other than Pakistanis.

The phenomenon of residential clustering can also be seen as a response to experience and cultural evaluations. A measure of residential concentration would have developed as a natural concomitant of the attraction of inner-city property but two other forces seem to have strengthened this tendency: firstly, the existence of chain migration, where early migrants were expected to provide accommodation for later arrivals; and secondly, a conscious effort by South Asians to retain propinquity when purchasing property. The spatial impact of these two associated forces is illustrated well in Anwar's (1979) study of the Pakistani population of Rochdale. He notes how one particular house in Tweedale Street had acted as a port of entry for a substantial proportion of the town's Punjabi population and how this property became the focus around which other Punjabis settled. Houses in other districts of the town acted as similar poles of attraction for different groups, such as the Mirpuris. In fact, Anwar and Dahya (1974) have gone further than this and shown how it is not just different regional-linguistic groups but even particular villages in India or Pakistan which have their own ethnic enclaves in British cities. The social outcome of this finely differentiated ethnic clustering is to recreate Indian or Pakistani village life within the British inner city. The development of ethnic territory thus fulfils a clear function in that it provides the security, stability, and support of the village and allows everyday life to take place within an environment which is value-reinforcing rather than value-challenging. It also re-establishes meaningful social sanctions against those deviants who show a desire to transgress behavioural norms. It is advantages such as these which stimulate Indians and Pakistanis to remain voluntarily clustered even after the cessation of chain migration, or even during residential mobility (see Werbner, 1979) and it is also these positive features which are recorded by housing and neighbourhood preference surveys.

Residential concentration also encourages the development of ethnic businesses since it offers a compact, dense, and partisan, potential market. The entrepreneur provides an additional way for his customers to avoid interaction with British society and, in exchange, he receives trade which ensures his survival in a market niche where a white competitor would fail. Ethnic businesses thus simultaneously fulfil Boal's (1978) 'preservation' and economic 'attack' functions. In addition, the developing literature points to the way in which the ethnic business sector provides employment for fellow ethnics, a channel for gossip and communication, a daily reminder of ethnicity, and a way to retain earned income within the community (see Desai, 1963; Forester, 1978; Cater and Jones, 1978; Cater et al., 1979; Mullins, 1979; Aldrich, 1980; Mullins, 1980; Werbner, 1980; Aldrich et al., 1981; Baker, 1982; Jones, 1982; and Robinson and Flintoff, 1982).

Those South Asians who were unable to find employment within the community in fields such as retailing were forced into contact with British society in order to survive economically. Even in the work-place though, individuals did not have to sacrifice their cultural preferences for avoidance. Research on this area is not plentiful but Wallman's (1979) book *Ethnicity at Work* contains two chapters specifically concerned with South Asians. Brooks and Singh's (1979b) chapter is the more relevant here, for they describe how a system of brokerage develops in which the Asian broker acts as a middleman between his fellow countrymen and the white 'management'. The broker offers understaffed white foremen a steady supply of reliable workers, often from his own *ilaqa* (group of villages) and *patti* (sub-area of a village), and in exchange he receives the status of being able to find work for relatives, friends, and new migrants (Wright, 1964). The net result of the brokerage system is a process of chain recruitment and ethnic work gangs (see Aurora, 1967; Marsh, 1967; Wright, 1968; Patterson, 1968; John, 1969; and the Commission on Industrial Relations, 1970, for further examples of this tendency), the importance of this being that ethnically homogeneous workgangs produce very little social integration between Asians and whites. Again it seems that cultural preferences encourage South Asians to seek, or at least not refuse, a culturally reinforcing environment.

Similar but less broadly researched conclusions have come from work on other areas of Indian and Pakistani social interaction. Several authors, for example, comment on the importance of the Mosque, Temple, or Gurudwara as centres not only of religious life but also of social life and learning. More than this, though, they are central to the maintenance of ethnic boundaries. Other authors comment on the exclusivity of South Asian recreational activities: all-Indian badminton teams; all-Pakistani soccer or cricket teams; Indian films and cinemas; record shops selling the latest film scores from Bombay; and festivals of Indian dance and art. Each of these provides a parallel to forms of entertainment which already exist for the indigenous population but which are specifically created by the South Asian population along traditional or pseudo-traditional lines and based upon ethnic identity. Lastly Rex and Moore (1967) and John (1969) have respectively discussed the roles of the Pakistani Welfare Association and the Indian Workers' Association. John, in particular, details the way in which the IWAs strongly paralleled the role of the traditional village *panchayat* (or council) and were made up of alliances closely resembling the village *partis*. For John, then, the purposes, institutions, and processes of village politics have been transferred in a mildly modified form to parallels in the context of life in Britain. The provision of such a parallel political system not only provides a power base for influential South Asians, but also removes the need for contact with the wider society.

In sum then, the interaction of Indian and Pakistani migrants in Britain is controlled by norms, sanctions, and values which are derived directly from those currently prevailing within the village of origin. Migrants seek to create an environment which will strengthen, rather than challenge, these values. They do this by restricting interaction with individuals who are members of the indigenous population in the neighbourhood of residence, and in the work-place. They construct parallel economic, institutional, political, and leisure structures which replace those of the indigenous society. These parallel structures, moreover, take the form and aims of those in the village. Activities within these structures are also conducted according to guidelines based closely on those of the sending village.

Intra-Asian Avoidance

All too often, researchers assume that the cultural forces act only to limit the contact which takes place between South Asians and whites. They ignore the fact that the South Asian population in Britain is made up of a series of mutually exclusive demographic groups, each of which is capable of developing its own independent way of life with its own institutions and value-system. In theory, these groups could be so different and so independent that they share little except a common colour and the same structural position in British society. Given this, it is likely that cultural forces will operate to shape and constrain the patterns of interaction which exist between different groups within the Asian population, as well as between Asian and white.

Criteria for fission within the Asian population exist at a number of different levels each of which overlaps to produce a mosaic of sub-groups. The coarsest, but one of the most important of these, is whether a group has had experience of life in East Africa or has migrated direct from the sub-continent to Britain. The former are described as being more 'westernized' or 'urbanized' (Lyon, 1972), and possess a better education, higher occupational status, and greater wealth. According to Thompson (1970) these perceived differences have generated stereotypes within the Asian population. Thompson's sample of East Africans characterized Punjabis as being vulgar and coarse; whilst the Sikhs thought of the East Africans as being aloof and clannish. Jenkins' (1971) data support the existence of these stereotypes in Leamington Spa. However, the importance of these views is that they have direct behavioural correlates. Thompson (1970) shows how they infect sport, seating arrangements at an Asian theatre production, and the choice of school friends. Michaelson (1979) demonstrates their salience for the arranged marriage system, even within one Gujurati caste. And Husain (1975) and Phillips (1981) discuss the spatial separation which has developed between the groups in Nottingham and Leicester.

A second broad criterion of fission is that of rural and urban origins. Khan (1977) has commented upon this within the Pakistani Muslims of Bradford, where urban or rural origins are clearly associated with sophistication and education.

In terms of its ability to unite apparently dissimilar groups, the axis of nationality, and its associated force of patriotism, appears to be the next most coarse differentiator. Jeffery (1976) mentions the fierce patriotism which exists amongst Pakistanis, but Anwar's (1979) more recent account is illuminating as to the strength of emotion which is involved. He writes (1979: 181):

The Indo-Pakistani wars of 1965 and 1971 led most Pakistanis both in Pakistan and in Britain to regard Indians as their enemies. In 1971 tensions arose in this country among Pakistanis, Bangladeshis and Indians over the Bangladesh issue. There were marches and counter-marches at both local and national levels in support of government actions back home; collections for defence funds were made and national organizations . . . were formed. At the local level special committees were formed to collect funds and to deal with the Press . . .

Bentley (1972: 46) underlined Anwar's account of events when he described the local reaction in Bradford to the 1971 war:

Much less emphasis was placed upon welfare matters (by the voluntary associations) and greater energy was directed towards mobilising Bengalis in mass rallies, pickets against the touring cricketers from Pakistan, and fund raising as part of the campaign to press for the recognition by Britain of the State of Bangladesh. During the war, nevertheless, relations between West Pakistanis and Bengalis in Bradford only rarely deteriorated to the point where violence took place.

Dahya (1973: 244) shows how such emotions became translated into patterns of social and spatial interaction: 'the main division of community is between those from Bangladesh and those from Pakistan. At the level of interpersonal relations, there is no interaction, and no Bangladeshi lives with those from Pakistan.'

Religion is often thought of as one of the most powerful sources shaping intra-Asian relations but, despite this, its spatial impact has not often been researched. Jeffery (1976) considers it to be the crucial element which prevents a unified Pakistani community developing in Bristol. She notes (1976: 157) of the Christians, for instance, that 'some of the men work with Muslims, but the ties are not brought into their homes and the women have little contact with Muslim people. Muslims are generally considered unsuitable contacts.' She also argues that it is the question of religion, coupled with national sentiments, that divides the Pakistanis and Indians. She observed (1976: 146) that 'apart

from sharing shops and cinema showings, there is little other contact between
Pakistanis and Indians in Bristol. In some respects this is surprising, for in most
matters . . . the West Pakistanis and Indians in Bristol are culturally and
linguistically similar'. Rose (1969) has also commented upon the way in which
religion may act as a powerful force in creating and maintaining communal
solidarity but his comments concerned the Sikhs. He wrote (1969: 452):

Their cohesion is not solely an affair of group settlement of fellow villagers (mostly from
the Punjab area of India), but springs from their common religious faith and from their
sense of belonging to a brotherhood which was forged in persecution.

Both Helweg (1979) and Lyon (1972: 6) agree, since for the Sikhs

religious ties are central and exclude fellow regionals in Britain. Sikhs . . . rejecting
Hindu idolatory, caste inequality, and priests, became obliged to fight Muslims for
their minority identity. Religion has ever since remained the crux of the militant
communities, so that in Britain the collective bonds of Sikhdom exclude Hindus and
especially Muslims; even if the latter come from Old Punjab and share the *lingua franca*
of Punjabi and the regional culture, they are separated by traditional enmities and
contemporary state boundaries.

Recent events in and around Amritsar have proved this point conclusively.

Beyond the issue of religious faith there is also the cross-cutting question of
regional identity which may serve to divide national or religious collectivities.
Shah (1979) notes how this has occurred within the Jain religion in East Africa
and more recently in Britain. She describes how there are two main groups
within the Jains, the Visa Oshwals (or Halari Jains) and the Cutchi Jains, one
of which originates from Saurashtra whilst the other comes from Gujarat. Shah
argues that, despite a certain co-operation and common sense of unity between
these different regional adherents to the same religion, individuals
predominantly think of themselves as part of a regional collectivity rather than
a shared religion. Kanitkar (1972) also indicates the unifying influence of
regional origin and its social value amongst the Indian élite in Britain. She
describes the way in which regional allegiances are activated in time of need,
particularly when one member of the group is searching for accommodation.
This frequently results in the residential clustering of co-regionalists in Britain
either in the same neighbourhood or lodging house. Bentley (1972) takes up the
same point when he notes the residential concentration of Gujuratis and
Punjabis in Bradford.

However, whilst most authors consider that regional origin is an important
basis for group solidarity, the majority also emphasize the role of a shared
language and therefore prefer to think in terms of regional-linguistic
groupings. Whilst these have been frequently mentioned in the literature,
credit must go to Desai (1963) for being one of the first to explore their implications for patterns of Asian interaction and residence in Britain. His work was
farsighted, especially when one considers that he was writing at a time when
society and academia were concerned with generalities such as the 'colour
problem'. He began by describing the various religious and linguistic characteristics of each of the major Asian groups who had migrated to Britain. At the

time, this in itself was new information to most interested commentators. However Desai went on to discuss some aspects of the patterns of social inter-action which existed between these groups. He wrote (1963: 18):

It is all the Gujuratis or all the Punjabis who constitute communities. The Muslims within the Gujurati community behave as a Gujurati caste. The same is true of Pakistanis, for East Bengali Muslims do not identify themselves with the Punjabi and the Azad Kashmiri. The three groups from Pakistan are isolated from one another. If anything, the Punjabi Muslims from Pakistan enter into inter-personal relationships more easily with the Punjabi-speaking Sikhs and Hindus from India, than they do with the East Bengalis.

These linguistic-regional groups are reinforced by cultural values and social relation-ships brought over from India, and are distinctive for each group. These discourage the individual from closer contact with a person from a different group.

Desai then elaborated on this last point and remarked on how a Gujurati would choose to live, work, and spend his leisure time with another Gujurati in preference to a member of one of the other groups. He also commented on the fact that this might have a direct spatial impact upon patterns of residence. Although he was quick to point out that whilst not all of these regional-linguistic groups were residentially clustered, many were. In particular, he cited the case of the Bradford Gujuratis who were highly spatially concentrated and who could, therefore, support a Gujurati cultural association, a Gujurati library, and a Gujurati club. Even where members of a regional-linguistic group found themselves scattered in different towns or regions within Britain, Desai described how this need not be an impediment to social interaction and group consciousness because of the willingness to maintain contacts through perpetual visiting at weekends and holiday times.

Desai's early work on the Gujuratis has since stimulated a good deal more research on this group, much of which argues that the common language and regional culture which the Gujuratis share override the divisions which exist within the group on grounds of religion or caste. Gujurati Muslims, for example, are simply treated by Hindus as a separate caste within the local hierarchy. Hahlo (1980) has also taken up this theme and maintains that the common regional origin of the Gujurati Hindus and Muslims, their youth-fulness as migrant groups, their common language, their low educational qualifications, and their tendency to cluster in the same ethnic areas in British cities all provide sufficient shared characteristics for religion to be forgotten. However the Gujuratis have not been the only group which has figured in the British literature. Chansarkar (1973), for example, describes the unique Maharashtrian regional group with its Marathi language, and notes how the group's cohesiveness has become evident in its residential clustering in North London.

At a finer level of analysis, commentators have demonstrated the salience of caste (*varna*) or sub-caste (*jati*) as an axis of community fission. Michaelson (1979), in particular, showed that caste affiliation affected identity since it influenced factors such as occupation, marriage rules, physical appearance, accent, vocabulary, and methods of food preparation. This identity was also institutionalized in Britain by the growth of associations and organizations for each of the major sub-castes (except the Patidars), and by residential concentration. Ghuman (1980) has confirmed much of this in his study of the Bhattra Sikhs in Cardiff. Finally, Lyon (1972) underscores the salience of caste or sub-caste when he argues that a major element of the Gujurati and Sikh identities in Britain is produced by the fact that the migration of each of these groups has been dominated by one *jati* and that this, therefore, develops a false homogeneity. Michaelson (1979) feels compelled to disagree.

On a similar level to differentiation by caste, one may note that exclusivism can also be based on differing interpretations of the same religion and their perceived status. Hallam (1972) has thus described the exclusivism of the Ismaili Muslims in East Africa and, to a lesser extent, in Britain. In this case though, exlusivism arises not only because of contrasting interpretations as to which individuals (or Imams) represent the true inheritors of the spiritual and temporal attributes of the Prophet, but also because of marked secular inducements. Barot (1972) provides a second example of the fission which arises from religious sectarianism. He reports a study of the Swaminarayan Sidhanta Sajivan sect, a Gujurati Hindu group with branches in London and Bolton. The sect is characterized by its own moral and social code, its own dialect of Gujurati, and a complex organizational structure which incorporates a judicial body. Affiliation to the sect not only involves a lifestyle tailored to the achievement of *moksha* (or salvation) but also a commitment to help fellow members, and considerable financial contributions. From a territorial point of view, Barot (1972: 34) has this to say: 'Despite the hazards of the London housing market, they had succeeded in settling in one locality, thus giving the *mandal* a well defined territorial basis. In Bolton, sect members have established themselves in a central district . . . The community has a specific focus and the members live in close proximity to each other.'

Thompson introduced an even finer axis of differentiation. Writing of the Punjabi Sikhs in Coventry he noted that 'some sub-groups have settled in particular areas but these are groups of villagers from the same village in Doaba.' (Thompson, 1970: 28). Thompson was therefore suggesting that a further level of fission existed beneath that of Desai's regional-linguistic group, and that these village links were manifested in social and residential patterns. Dahya (1974) took up this theme at a later date when he proposed a chronology through which Asian settlements in Britain are passing. This can be summarized as follows:

The process of growth and subsequent development of the immigrant community is one of fusion of immigrants from different areas leading to a fission and segmentation on the basis of village-kin ties. That is, the process begins with the fusion of members of various ethnic/sectarian/national groups during which stage traditional attitudes of inter-ethnic/sectarian hostility are temporarily shelved. (Dahya, 1974: 87).

Dahya's more detailed description of this process underlined his support for Thompson's belief in village-kin as well as regional-linguistic segregation: 'fission and segmentation of the immigrant community from fortuitous sub-groups into sub-groups of choice is a form of developmental cycle where the immigrants exclude themselves from those outside their village-kin groups and express their preferences for living with those of their own kind' (Dahya, 1974: 87–8). For Dahya, then, each regional-linguistic group must be seen as a loose alliance of interlocking and interdependent village-kin groups which share common attitudes to their homeland and a common past.

Jeffery's (1973) doctoral research took the issue of village-kin cohesion one stage further, and additionally placed it upon a more solid base by defining more accurately the exact limits of the village-kin group. She wrote 'there is boundary maintenance within the Pakistani settlement, and the way in which it operates is very similar to the way the boundary maintenance between Pakistani and British people operates' (Jeffery, 1973: 213). The end product of this, she argued (p. 219), was that 'the settlement of Pakistanis in Bristol is highly fragmented, for there is a tendency for kinship clusters to develop which are relatively discrete'. She preferred to use the more precise term of *biraderi* rather than the alternative 'village-kin' concept. The former could be accurately delimited by the participants and was a concept peculiar to Pakistani society, whereas the latter was nebulous and had been imposed by western analysts. Jeffery thus preferred not to think of Pakistani settlements as communities but as a collection of mutually exclusive overseas satellites of Pakistani *biraderis*. The important distinctions between this view and that employing the notion of village-kin groupings, are that villages frequently contain several *biraderi* which may not associate overseas, and in addition, overseas *biraderi* may include substitute kin and friends defined as such only in the context of life in Britain. As well as making finer distinctions within the axes of fission, Jeffery also provided several examples of ethnic animosity at work within the Pakistani population of Bristol.

It seems then that Indian and Pakistani migrants in Britain not only avoid contact with the indigenous population, but also minimize the need for interaction with other Asians whom they feel to be members of out-groups (on whatever criteria appear appropriate at that time). They employ similar strategies for both purposes and these strategies produce similar spatial, social, and institutional outcomes. The desire to maintain cultural purity in order to facilitate a ready re-adoption into the village on return migration suggests itself as one of the most likely causes of this behaviour.

The Maintenance of Links with the Sending Society

Whilst abroad, the Indian and Pakistani migrants thus make very strenuous efforts to recreate both a cultural environment and a social structure which are similar, in all their major dimensions, to those of the sending society. By doing so, they emphasize and defend their identity and also create a situation where they need only occasionally enter into positions which are culturally threatening. In addition to this, many migrants underline their transience and their determination to retain their culture and ways of life by maintaining physical, identificational, and financial links with the society from which they migrated.

These links take a number of forms but they all act as persistent reminders that the migrant is still a permanent, but temporarily absent, member of the village-kin network or *biraderi*. The migrant is thus prevented from forgetting or ignoring his identity and the obligations which are associated with that identity.

Perhaps the most direct form which such links may take is demographic. The migrant may be despatched by the joint household to gain an additional source of income, but he is frequently required to leave his wife and children in the sub-continent. Thompson (1970: 88) suggests reasons why this should be so: 'the absence of women at this stage of the migration reflects either the short-term intentions of the migrants or an apprehension of commitment to settlement in a new environment'. Dahya (1974) is more pragmatic about reasons for male-only migration. He argues that the financial cost of arranging for the men to be joined by their wives could not be justified on the ground of potential economic return. Moreover, he suggests elsewhere (Dahya, 1973) that the head of the joint household might deliberately retain the migrant's wife and children in order to ensure a steady flow of money from overseas to offset the initial 'investment' in the migrant's air fare. Dahya went on to argue that, even where wives were allowed to leave the sub-continent, female children were still often left behind, and therefore the link with the sending society could not be broken. Jeffery (1973) presents similar reasoning but with regard to aged relatives who rarely accompanied Muslim migrants overseas. However it is important to qualify such comments by noting that the Muslim migration referred to in the works above has been the most demographically unbalanced of all the Indian and Pakistani migrations, and therefore the comments of Dahya and Jeffery probably do not apply with equal force to other groups. Thompson (1970) does note however, that there is a similar tendency amongst Sikhs to leave daughters in the care of the joint family in preference to having them come to Britain. Leaving members of the joint family behind in the sub-continent thus acts as an important force in reminding the migrant that he is still part of the village and is expected to behave accordingly.

The second link which the migrant retains concerns his own presence in the village. Brooks and Singh (1979a) do not regard this as an accurate measure of the intensity of feeling towards the country or village of origin, but they do provide data on this: 71 per cent of Punjabi Sikhs had never visited their village of origin since coming to Britain but the percentage was considerably less for the Gujurati Hindus and even smaller for the Punjabi Muslims. The latter were also the most likely to make extended visits and sometimes gave up their jobs in order to do this. The Sikhs, in contrast, made only brief and functional visits. Jeffery (1973) has elaborated the role of visiting within the Muslim group by describing how, because of geographical dispersal, gift-giving has to replace hospitality. Annual visits by one member of the *biraderi* thus ensure that a courier is available to take presents (such as electrical goods) back to relatives in the village. In return, the migrants are sent examples of traditional handicrafts which remind them of their family and village. Messages are also transmitted between family members in this way. The exact timing of the visit may also be arranged to coincide with weddings or other family events or crises; this again emphasizes to the migrant that he is still regarded as, and should regard himself as, an integral part of that community.

A third link which the migrants maintain is that of marriage. Because of the arranged marriage system and the excess of Asian boys in Britain in comparison to the number of Asian girls, migrants remain tied both to their parents and to the sub-continent for a suitable match. The groom has to behave in Britain in a manner which will ensure that his chances of a good marriage are not diminished.

The fourth way in which migrants remain a part of the sending society, despite their physical absence from it, is through the remittance of money to the head of the joint family. Analysts frequently regard this as the best indicator of emotional links and it is also likely that it impinges upon everyday behaviour and life-styles more than any other factor. Dahya (1973) was one of the first to look at the topic and observed that the use to which remittances are put changes over time. Initially they are used to pay off any debts outstanding from the cost of emigration. They are then channelled into the procurement of additional land and its more intensive use through the provision of tube-wells. Later, attention turns to the construction of a *pukka* house on the outskirts of the village, and finally to investment in a business such as a flour mill or shop. However the main issue is that decisions about the disposal of remittances are not taken by the migrant himself, but by the head of the joint family. One factor which will weigh in such decisions is the head of household's desire to ensure the eventual return of the migrant to the village. Recently, Helweg (1983) has undertaken more detailed research in this field although the literature still does not contain a body of knowledge which can match the contributions of such authors as Manners (1965), Philpott (1968; 1977), or Frucht (1968), each of whom has studied the importance of remittances from Britain to local economies and social structures in the Caribbean.

Another area which appears to be under-researched in comparison with its potential significance is that of language. Robinson (1980c; 1982b) has discussed aspects of this in relation to the learning of English and as an indicator of identificational assimilation amongst East African Asians and South Asians in Britain. However there is a need for much broader analyses than these, possibly relying upon the overlap between social geography, sociology, and socio-linguistics (Trudgill, 1974 provides an introduction to this topic by discussing the relationship between language and social distance). Until more detailed work is undertaken, one can only outline in the barest of detail the role of mother tongues as cultural markers and ways of deliberately maintaining interactional boundaries (Barth, 1969). In the meantime, both Thompson (1970; 1974) and Jeffery (1973) have commented on certain aspects of the topic. The latter noted how hardly any of her female respondents spoke English regularly and how even the men, who had to learn English to survive in the work-place, were not comfortable with the language and preferred to use their mother tongue where possible. Thompson (1970) was more concerned with the issue of youth, and in particular the younger Punjabi Sikhs. He, like Jeffery, found that Punjabi was the main language used within the house and that this was taught to children as their first language prior to them going to school. He also observed that the organization of peer groups was such that Punjabi youths always associated with other Punjabis and the language of their activities was again Punjabi. This, he argued, had important repercussions since 'fluency in Punjabi allows the young people to enter fully into first generation situations, such as political meetings, discussions in the pub, and family gatherings. Full participation in such activities itself reinforces the individual's Punjabi identity and helps to maintain and develop the language, which itself is continuously in use in such situations.' (Thompson, 1970: 250.)

Thompson's mention of political meetings introduces another area where migrants can express the strength of their commitment to their homeland. John's (1969) book demonstrates one facet of this in the form of Indian Workers' Associations, but he argues that these were largely orientated towards the immigrant situation in England. Thompson (1970) on the other hand provides a clear example of political action amongst Sikhs in England which was directed towards the home society; this was the *Jai Hind* party in Coventry established during the 1960s by local young Punjabis. Although *Jai Hind* had no connection with any existing party in India and was a product of circumstances in Coventry, its focus of attention was India, and more specifically events in the Punjab. It sought to bring revolution to the Punjab from its English base. A further example of political continuity between the home society and the receiving society is provided by Marsh's (1967) account of the strike at Woolf's rubber plant in Southall in the mid-1960s. There, animosities between Indian and Pakistani groups which pre-dated migration surfaced in trade union organization and industrial action.

The final link which the migrant retains with the sending village is the obligation to help kinsmen on their arrival in Britain. Thompson (1970) at least, regards this sponsorship of passages and encouragement of chain migraton as *the* most important of all the links. Not only does it affirm that the overseas migrant still takes his obligation towards other village members seriously, but, by ensuring that he is continually in contact with new migrants direct from the village, it acts as a crucial brake on any tendencies towards cultural aberration on the part of the established migrants.

The Myth of Return?

The motives which encouraged the migrant to leave the village in the first instance, his preconceived attitudes towards the value of British society and its peoples, the perception that residence overseas is only a temporary phase, the support and protection afforded by the fledgling village-kin group in Britain, and the continued links which the migrant retains with the sending society, all encourage South Asians to retain and defend their culture, identity, and ways of life whilst in Britain. This is predicated upon a belief that 'eventually' they will return 'home' to retire and enjoy their new status. Whilst the strength of this phenomenon varies between the different Asian regional-linguistic groups, it appears to be a part, however small, of each of their ideologies. But, the critical question remains to be answered. Is the premise of return a myth, or a commitment?

It is only necessary to point to those factors discussed in the previous section to indicate some of the major forces stimulating return migration. Migrants are encouraged to believe they are still an integral part of the village and the joint household. They recreate traditional ways of life and institutions in Britain and they maintain continued physical and emotional links with the village of origin. Their savings are invested in the homeland and the majority of their relatives still live there. Life in Britain must not, therefore, appear to offer a serious alternative to an eventual return 'home'. Coupled with these strong attractions are the 'push' forces such as prejudice, discrimination, the activities of right-wing organizations, and economic recession in the UK (see Selbourne, 1983a, 1983b for the effect of this on West Indian return migration).

The forces which discourage return need to be discussed in greater detail and fall conveniently into three categories: changes in the sending society; changes in Britain; and changes internal to the migrant or migrant group. Jeffery (1976) considers one of the most important of the changes in the sending society to be the relative rates of inflation in India, Pakistan, and Britain. She argues that because inflation is so much higher in Pakistan than in Britain the migrant is continually forced to delay his return trip because his savings become insufficient to ensure a return to the higher life-style and status which were the reason for leaving the village. Connected with this is the parallel question of dowries. Tambs-Lyche (1980) states that the dowry expected from a *vilayati* (returning migrant) has also undergone inflation (largely because of the exaggerated stories which filter back to the village about the wealth available in Britain) and the migrant therefore has to delay his return in order to save more money to attract a wife than would have been the case prior to migration. The mere fact of having migrated thus extends the period over which the migrant must remain overseas in order to achieve the same goals. For Khan (1977) though, a more potent force limiting return migration is the *absence* of economic change in the sending society. She states (1977: 68) 'many of the conditions which led villagers to emigrate also explain the notable absence of large numbers of permanently returned migrants'. In contrast, Dahya (1973) concentrates upon the plight of the returnees, and how they are forced to play the role of wealthy adventurers who have become westernized whereas they wish only to return to the traditional way of life. Moreover, many migrants forget over a period of time that the sending society had undesirable characteristics such as the prevalence of corruption or abject poverty. Stories of *vilayati* who have returned and been shocked by such forgotten characteristics must inevitably discourage others from following their lead. Finally Dahya points to the fact that the migration itself (and the remittances which formed an important part of this) may have been instrumental in destroying many of the features of village life which were cherished by the migrant whilst overseas (cf. Davison, 1968 and H.O. Patterson, 1968 on returning West Indians).

Return migration may also be discouraged by changes which take place in

Britain, most notably those which relate to government decisions on immigration. Both Taylor (1976) and Lawrence (1974) have commented upon this. They suggest that, in the light of the progressive tightening of the criteria for entry to Britain, many South Asians remain in Britain who perhaps would not ordinarily do so. Migrants who are unsure about the wisdom of returning to the village are unlikely to risk this course of action since they know that it would be extremely difficult to get back into Britain if their stay were to prove unsatisfactory.

Finally, there are those changes which are internal either to the individual migrant or to the entire group. These may relate to either financial, social, or attitudinal matters. Lawrence (1974) provides the best discussion of the former. He notes that some Asians are unable to return to the sending society for the simple reason that they cannot afford to do so. A more frequent situation is where the migrant *could* afford the cost of travel but would not then possess sufficient additional savings to adopt a life-style on his return which corresponds to that expected of a successful *vilayati*. The migrant thus chooses to stop in Britain for a longer period to allow the accumulation of more funds. This, however, may create further difficulties since the family of the migrant will raise their expectations of his wealth with each successive year of his absence in Britain. The Indian or Pakistani is thus trapped in a continuous cycle from which he is unable to escape without considerable loss of prestige. Finally, Lawrence (1974) argues that analysts are occasionally much clearer about the financial targets which stimulated migration than the migrants are themselves. For many of the latter, targets change and become blurred so that there is rarely a logical time for return.

Social changes which the migrant himself experiences may also influence the decision whether to stop in Britain or return to the village. In particular, Jeffery (1976) suggests that some Asians in Britain have realized that despite their newly acquired wealth they could not hope to return to a truly prestigious position because they are uneducated. Since education is not easily attained later in life, many have instead consigned themselves to enjoy material satisfaction and the status which can be achieved from this, within the social arena of life in Britain. Moreover the feeling of blocked social mobility which uneducated parents have experienced, seems to have intensified their desire to ensure that their children do not suffer in the same way. Because of this, some parents have adjusted their aspirations away from the accumulation of money and a speedy return to the village, towards a longer-term strategy involving the education of their children to the highest level available to them in Britain. This has inevitably postponed return migration. In a similar fashion, the very achievement of education by the second generation may itself be postponing or cancelling any thoughts of leaving Britain. Once children are educated to 'A' level or university standard, and have gained responsible or professional employment in Britain, the attraction for them of returning to farm their father's land in the village might be slight.

The last factor which might delay return migration is those changes which occur in the aspirations of the migrant as a direct result of life in Britain. What might have seemed to be luxuries in Pakistan (such as a car) come to be regarded as necessities in Britain, and the migrant feels it essential to stop longer here in order to be able to take such items back with him on return.

The second generation

Most, if not all, of what we have said so far relates to those first generation migrants who came to Britain of their own accord either as married men or women, or as single men. What is becoming of increasing importance with the passage of time is the attitude of the second generation (and indeed the third generation) who either came to Britain as child migrants or were born here. The attitudes and aspirations of these people will not only determine whether they themselves remain permanently in Britain but may well indirectly influence the behaviour of their parents.

Several studies exist which concentrate specifically on the question of second generation adherence to the norms of their parents. Their conclusions are similar despite the fact that they consider different regional-linguistic groups. However, one must note that none of the surveys was undertaken very recently and their findings might not therefore be representative of current thinking. Again this is an area in urgent need of further contemporary research. Taylor's (1976) study of Sikh and Muslim youths in Newcastle does, however, shed some light on the issue. He found that when asked where they would settle in the long-term, 30 per cent of youths said the UK, 37 per cent said they would return home and the remainder were undecided. Taylor, sensibly did not interpret these results at face value but concluded that Asian youth formed a 'half-way generation' who were no longer committed to returning home but who still adhered to traditional ways, and still regarded themselves as being apart from British society and British people. Taylor went on to support this claim by showing that, whilst Asian youths had been educated and brought up in Britain, their parents still formed the dominant reference group and few thought that traditional ways should be radically altered.

Brah (1978) considered a different aspect of second generation conformity. She looked at the attitudes of a sample of South Asian teenagers in Southall towards marriage and ethnic identity. Her enquiries revealed that, despite pressures in the opposite direction, most teenagers accepted arranged marriages and the implied conformity which went with these. Their attitudes were however, somewhat different on the issue of courting, where the teenagers were happy to form temporary relationships with members of other religions or castes. Brah thus argued that whilst the racial divide remained intact within the sphere of marriage, internal divides between different regional-linguistic groups were increasingly coming under pressure. The decline of the latter might be seen as a tentative move towards more widespread intermarriage with the host population as predicted by such authors as Bagley (1972). Robinson (1980d) is not able to agree.

Perhaps the most thorough piece of research on inter-generational differences within the Asian population is that of Thompson (1970; 1974), who studied the Punjabi Sikhs of Coventry. Thompson was especially interested in the transmission of cultures and ideologies between Punjabi parents and their children, and the points at which such inter-generational communication might break down. He began by drawing the important distinction between child migrants and second generation children born in Britain. However, before discussing the way in which these two groups reacted to life in England, he outlined the common forces which encouraged both to accept, and conform to, traditional ways. Central amongst these is the joint household, and the common life-style and purpose which this imposes upon all its members. Such consensus is manifested in both the economic and social domains. With regard to the former, Thompson (1974: 246) wrote:

> The regular handing-over of their wage packet binds the 'workers' closely to the joint family, asserting and reiterating their membership of it. As contributors from an early age to the joint family income, these boys' economic futures are caught up with the economic future of the family, which is probably in India. If the son's contributions are channelled to buy land and so on, he will be orientated to a future in India for, besides being the son of his father, the fruits of his labours have gone into the family holding there.

However, the joint family and the loyalty of the second generation towards it, not only ensures that the *economic* future of its younger members is inextricably bound up with that of the first generation but it also imposes considerable social demands. In effect, a commitment to the joint family is also a commitment to behave according to norms familiar in the sending society, for without this the family's *izzet* would be under continual pressure. Nowhere is this more so than with the question of marriage. Agreement to an arranged marriage may well allow the family to gain considerable prestige, whilst refusal to do so almost inevitably causes loss of face. It is hardly surprising therefore that Thompson is able to comment that the major change between first and second generation marriages is not *whether* they are arranged but *how* they are arranged.

Another common mechanism by which the joint family enforces conformity on child migrants and second generation children alike 'is by sending them to visit the village in India. These visits are more than sentimental journeys. To visit the Punjab is to go "home", to the place which the elder members of the family have always regarded as their home and to which they intend to return. It is also the place where money earned in England has been invested in land, buildings, or a small business.' (Thompson, 1974: 246.) Whilst there, the child is treated lovingly and generously by his relatives and he also enjoys the prestige of his status as a *vilayati*. Frequently such visits lead a child to adopt the ambitions of his father, i.e. to return 'home' after deriving maximum benefit from life abroad.

Finally, the spatial and social segregation created by the first generation specifically to avoid sources of cultural threat has the effect of denying the Asian child the opportunity to choose or mould his own identity from a number of alternatives. The same could also be said about the maintenance of the mother tongue.

Whilst Thompson indicates that there are these common forces encouraging conformity to first generation norms, he describes how the opportunities presented to the child migrant and the second generation child differ, as do their responses to these. The child migrants overwhelmingly marry spouses from India because of their desire to retain links with kin and friends at 'home' and because of their idealization of village life and the personality traits it encourages. In contrast, the second generation boys are not nearly so uniform in their choice or their acquiescence to their parents' choice. Some choose to marry a spouse from the village, but increasingly contacts with other Asians in Britain are used to arrange matches between Asian boys and girls born here. Thompson also cites cases of where second generation children brought pressure to bear on their parents to 'arrange' a match which would formalize what was already a love-match. These trends will, of course, diminish the orientation of the younger Asians towards the sub-continent and will instead begin to create a separate marriage pool within Britain. This clearly has implications for cultural deviance since it would remove the continued input from the village and therefore shift the reference group against which British Asians compare themselves.

The migrant and second generation children in Coventry also differ in their attitudes to work. The latter were more often engaged in full-time education or apprenticeships and, as a result of their lower incomes, were not expected to contribute as much to the finances of the joint household. This reduced their commitment to India and to the family's landholding there. In addition, they had a more British view of work and sought only an eight hour day, thereby providing ample opportunities for recreation and leisure. Despite these facts, Thompson considered that the responsibilities which these boys felt towards their families would ultimately ensure that they continued to behave as Punjabis and joint family members. The same could also be said of the child migrants but for different reasons. Their limited ability to speak and write English and their lack of formal educational qualifications threw them back onto the values of the first generation and a commitment to return migration.

Thompson isolates attendance at school as the second potential area which could challenge the transmission of culture. But, in Coventry, he discovered that both migrant and second generation children belonged only to Punjabi peer groups. Friendship patterns thus strengthened a Punjabi identity, although the exact membership of the groups proved to be a product of the immigrant situation rather than being based on the village of origin. Again, it seems that links are multiplying within the migrant community which could eventually herald the development of a Punjabi identity which is unique to members of the group who are resident in Britain.

Thompson concluded his thesis by arguing that the child migrants had simply been transported from a Punjabi society in India to one in Britain. The second generation had been more influenced by the culture of the corresponding cohort in English society, and had internalized much of this, at least at the superficial level. Despite this 'they respond enthusiastically to symbols of Indian and particularly Punjabi grouping. In their relationships with their associates they show a strong preference for Punjabi friends (and a Punjabi wife), and furthermore, they keep the company they prefer.' (Thompson, 1970: 356.)

Finally, and more recently than Thompson's study, the Community Relations Commission (1976) has undertaken research into the generation gap which it considers is developing within the Asian population. Although its sample was reasonably large and its data were detailed, the report itself was disappointing. Although it contained references to the 'stress and conflict' which had developed between Asian parents and youths, many of the data provided evidence of considerable, if declining, conformity: 80 per cent of both parents and youths supported the maintenance of the joint family: only 35 per cent of youths felt they wanted more freedom; and only 5 per cent of youths would be prepared to pursue a course of action that might damage the prestige of the family. The report did point to areas where Asian young people felt they might wish to diverge from what is thought traditional within the Asian population (areas such as wearing western clothes, dating, and leisure pursuits) but disagreements with parents on such issues are not unusual within the white population as youths undergo the transition to adulthood.

It would be unwise to suggest that conclusions derived from only four studies could be more than suggestive. This is even more true when these studies rely only on small-scale sample surveys in specific localities. Thus, we can only comment that on the basis of these findings there is a good deal of congruence between parents and children in their adherence to traditional norms and in their desire to maintain the culture of their homelands. This congruence is especially prevalent when one considers those children who were born in India or Pakistan and came to Britain with their parents. They differ from their parents hardly at all in their lifestyles and attitudes, and they seem to be equally committed to the concept of returning 'home' when they have achieved the target set by the joint family. Those Asians born in Britain are more problematic. Because of the influence of the joint household, continual visits to the village of origin, and restricted opportunities for out-group interaction they have internalized much of their parents' culture and outlook. Having said that, though, they differ from the elder members of their community in a significant manner; their occupational structure and attitudes to frugality do not suggest a speedy or successful return to the sub-continent; their own culture has developed in the immigrant situation and therefore contains elements of traditional and indigenous cultures; and they demonstrate that links are forming between different groups (village-kin at the moment but possibly regional-linguistic later) in Britain at the expense of ties back to the homeland. This suggests that in the long-term the second generation might abandon the idea of returning home because few of them desire it and even fewer could successfully make the transition. Instead they would retain certain distinctive elements of their culture which would mark them off from the indigenous population (and to a lesser degree from other Asian groups) but these elements would become detached from events and trends within the sub-continent and they might also lose the significance that they enjoyed there. Further into the future, structural forces could stimulate a resurgence of ethnic pride, a renewed interest in the sub-continent, and a desire to return there. Rastafarianism has provided a model for this. In the short-term, though, Asian youths remain locked into a life-style and orientation created by their parents as a defence against cultural invasion. Moreover, even if the will were there, the second generation currently lacks the numbers, power, or social influence to prevail against the attitudes of the first generation.

Empirical evidence

Whilst the attitudes of the second and subsequent generations may conceivably alter the long-term objectives of the population of Indian and Pakistani ethnic origin in Britain, this is unlikely to occur for some time. However, disregarding potential sources of change, it still remains important to discover whether the concept of return is a genuine commitment to do that, or whether it represents a convenient strategy to make life bearable in a racist society which rejects coloured migrants. Having discussed the forces which might be seen as encouraging or discouraging return, it remains to see whether any empirical data exist on the topic or whether any inferences can be made from other sources.

Considering the credence which has been given to the notion of the myth of return, and the potential long-term implications that it has for the shape and form of British society, there has been a surprising absence of detailed research on the topic. Data are sparse, ambiguous, and fragmentary, although it appears that this is not an unusual state of affairs regarding return migration (King, 1978).

One of the first groups to study the question was the Economist Intelligence Unit who published its findings as early as 1961 (EIU, 1961). For its survey it relied upon official immigration and emigration statistics which, at that time, were collected and published by the Board of Trade. It noted, however, that these data were not unflawed since they were derived from only a 50 per cent sample of ship manifests for sailings direct from non-European countries. They could therefore omit migrants who entered or left Britain either by air or by boat across the Channel. Despite these shortcomings the EIU proffered its opinion that the Board of Trade figures differed from the real figures by only a small percentage. The EIU reworked the data to show a percentage rate of return (i.e. emigration as a percentage of immigration) of 39 for the Indians, 63 for Pakistanis and 21 for British West Indians. The time period to which this related was 1956 to 1959 for the Asian groups and 1955 to 1959 for the Caribbean groups.

In the light of more up-to-date information, the EIU (1962) revised its estimates, although still employing the same definition and methods. Thus over the period 1956 to 1960 (or 1955 to 1960) return migration was calculated to have been 40 per cent for Indians, 69 per cent for Pakistanis and 19 per cent for West Indians. Although the EIU did not itself do this, the data, and the conclusions based upon them, are in need of some qualification. The Board of

Trade defined the ethnicity of migrants by reference to the country of which they were passport holders, not by the more usual British definition of birth-place. The Board of Trade did eventually standardize on the latter definition (from November 1960 onwards) but since it ceased to collect immigration data shortly afterwards, this change was of little benefit. Secondly, the statistics upon which the EIU's conclusions were based do not allow the calculation of rates of return as such. By separately measuring the number of immigrants and emigrants they tell us only the speed with which individuals circulated through the cycle of working overseas for a period of time before being replaced by another family member. They cannot tell us how many families sent sons overseas, achieved their aims, and were therefore able to withdraw from the migration altogether.

However, problems with the statistical base from which return migration rates might be calculated have not been confined to the Board of Trade data. The two major sources of immigration and emigration figures on which analysts must rely, are the Home Office and the International Passenger Survey (IPS). Neither of these data sets is without its problems and neither is sufficiently complete nor inspires enough confidence to encourage a detailed analysis of return migration. The Home Office data are collected under the terms of the 1971 Immigration Act and are therefore concerned solely with the number of individuals entering Britain and wishing to settle here. They are devoid of figures on emigration. The IPS *does* quantify both immigration and emigration but this is perhaps one of its only major advantages. It is financed by the Department of Trade and Industry specifically to provide information on tourism. It therefore covers all the principal air and sea routes and its sample size is large. However within this sample the overwhelming majority are not migrants but visitors: in 1975, for example, the total IPS sample was 117,000 but the number of immigrants within this was only 2773 of which a mere 44 were from the West Indies and 155 from India, Pakistan, or Ceylon. When the data were grossed up for publication the number of West Indian arrivals was declared to be 2900. Clearly there is a large margin for error here, and one which is unacceptable. In addition, though, the IPS varies the frequency with which sampling is carried out at different ports of entry for economic reasons, and this again generates inconsistencies. Finally, the IPS data are frequently revised *ex post facto* in order to prevent the discrepancies between these and the Home Office statistics becoming excessive. In the light of these problems, statements that 4000 Indians and Bangladeshis had returned 'home' (for a period of not less than twelve months) in the year ending 30 June 1983, in addition to 2000 returning Pakistanis (OPCS, 1984), carry little weight.

It is probably for reasons such as these that researchers have turned away from official statistics and have relied more upon limited surveys in an attempt to ascertain the intentions of Indians and Pakistanis in Britain. Again, how-ever, the limitations of small-scale, often popularized, surveys are such that conclusions must be seen as tentative. The EIU (1962) was again amongst the first to probe this topic using this technique. A professionally managed, national survey of 3000 immigrants in ten major cities revealed that 26 per cent of Indian respondents, and 40 per cent of Pakistanis were intending to leave Britain.

The Times (1968) reported a survey in September 1968 which was typical of the sporadic attempts by the media to quantify the permanence or impermanence of coloured settlement in Britain. It employed a sample of 420 respondents drawn from 20 different locations and concluded that 'more than a third of those questioned would like to return to their country of origin if they received financial help to do so'. It did, however, warn that its sample was unrepresentative and should therefore be treated with caution. Lord Radcliffe (1969) cited several similar or even more impressionistic conclusions where the desire to return home varied between 50 per cent (Jamaicans) and 100 per cent (Sikhs).

Although the Opinion Poll technique has persisted (despite the warnings of Brooks, 1969) the emphasis has tended to shift towards data being collected as part of a local community study. Lawrence (1974) has reviewed a number of these, and concluded that the limited evidence suggested a trend towards a greater orientation away from return and towards more permanent settlement. In many cases this finding related solely to West Indians. In Nottingham for example a mixed Asian-Caribbean sample evidenced a decline in intentions to return from 95 per cent to 75 per cent over a period of only a few years. Rex and Tomlinson's (1979) much more recent data now put this figure (for an Asian-only sample) at approximately 26 per cent. Although few of these studies have systematically collected data on the settlement intentions of an Asian population over time, the general, if vaguely calibrated, suggestion is that the percentage of Indians and Pakistanis who actually intend to return to the sub-continent has declined considerably since the early 1960s. Having said that, such suggestions are very tentative indeed, and it is important not to forget that a substantial number of South Asians *do* still profess to be committed to return some fifteen to twenty years after the peak of immigration.

An interpretation

Because of the sparsity and limitations of much of the data relating to the notion of the myth of return, a discussion of the matter cannot be conclusive. However, what is apparent is that the concept cannot, and must not, be simply interpreted at face value. The belief in return clearly has different meanings for different people depending upon their circumstances, psychological make-up, and aspirations. For some there is a genuine and real commitment to return migration either at a determinate or indeterminate time in the future. For others, the belief is a useful crutch which helps them survive in a hostile society: the knowledge that prejudice, discrimination, and deprivation must only be endured 'for a few more years' is both comforting and a means of making it bearable. For others the myth of return is a means of idealizing and preserving all that is thought best about the sending society and its culture. In other words it provides a reason why migrants who are under continual cultural threat should cling to their traditional ways. Finally it may represent a means of providing self-esteem for migrants rejected by the host society. The feeling that the British economy and society are simply being used to achieve goals set by the sending society might provide some sense of satisfaction in the face of continual accusations that it is the migrant who is passive in his role as part of an expendable replacement labour force.

What seems more certain is the fact that the balance between these motives has shifted significantly since the early period of mass migration, and that the myth of return probably now has a new meaning and importance. As Brooks and Singh (1979a: 19) indicate, 'the essence of our argument can be summed up in a sentence or two. It is that, whilst ''first generation'' Asian immigrants in Britain still see themselves as transients, imbued with . . . the ''Myth of Return'' to their native lands, their situation in the British economic and social structure, combined with developments in their countries of origin, will ensure that most do not return to those countries.' If the belief in return is no longer a solid commitment to a course of action why then has it persisted? Dahya (1973: 268) suggests a plausible reason: 'the migrant continues to re-affirm his adherence to the Myth of Return because for him to do otherwise would be tantamount to renouncing his membership of the village community and the village-kin group in Britain . . . The myth is an expression of one's intention to continue to remain a member of both of them.' He continues (p. 269), 'the myth is a means whereby the migrants are able to make a socio-psychological withdrawal from a commitment to the norms of the wider society'. In short, whilst the belief in return may no longer be as important in terms of its direct physical impact upon long-term settlement in Britain, it is increasingly important as a mentality which South Asians will carry with them regardless of the length of their stay in Britain. Probably of greater consequence for our purpose here is that this mentality will have far-reaching and long-lasting implications for the social behaviour of Asians in Britain, and this itself will be manifested in an overtly spatial form.

Summary

This chapter has shown how the forces built into Indian and Pakistani society and family life encourage an overriding concern with status and a compelling drive towards achievement. It has also shown how the joint family system transfers these ideals and responsibilities to the younger members of the household whose energies are frequently harnessed to achieve these familial goals. In the days prior to Empire, opportunities were available for social mobility, and aspirations were relatively restricted in the sense that they focused only upon landownership as a key to ritual mobility. The coming of the Empire changed all this. It reduced potential mobility but considerably broadened and intensified aspirations. It has been suggested that firstly rural–urban, and latterly international, migration was the outcome of this. In both these cases, though, the migrant had relatively modest 'targets' rooted in the needs of the joint household in the village. His absence from the village was regarded by all as only a temporary expedient. This feeling of transience coupled with a negative evaluation of British society and life-style did not stimulate the first generation immigrants to Britain to consider British culture as a viable or desirable alternative to their own. They consequently adopted strategies designed specifically to defend and reinforce their own culture. These involved minimizing social interaction with the indigenous population and maximizing links with the sending society. In short they developed a separate and self-contained satellite of their own society in Britain. All these factors combined to produce a myth of return amongst first generation Indian and Pakistani immigrants who declared themselves pledged to return migration once they had achieved their economic and social objectives in Britain. It has been argued, though, that this belief has undergone a change in meaning and significance of late, as more and more first generation Asians realize that they are either unable or unwilling to return to traditional ways of life in the village of origin. It is likely that this change will be intensified as the second generation gains positions of influence within the Asian population in Britain. Nevertheless, whilst the myth of return is unlikely to have any major short-term impact upon return migration, it is of considerable significance since it creates a mentality towards life in Britain which has far-reaching and long-lasting social and spatial concomitants.

However it is again vital to emphasize that the foregoing relates to first generation Indian and Pakistani immigrants and cannot, and must not, be extended to take in those Asians who came to Britain from East Africa, nor Christian Asians (see Jeffery, 1976 on this point). Chapter 3 described how the East Africans benefited economically and socially from the coming of Empire. They achieved considerable wealth in some cases, and came to Britain involuntarily rather than as a result of the country's economic attractions. The environmental conditions they enjoyed in East Africa would lead them to expect more of life in England than was the case for their South Asian counterparts. Finally, and most importantly, they had no myth of return, and, because of their experience as a middle class in Africa and the greater contact with overseas Britons which this had stimulated, their evaluation of British society was by no means as negative. This, and the absence of ties back to East Africa and the historical attenuation of links back to their ancestral homelands, did not produce such an introverted or isolationist mentality towards life in Britain. The East African Asian was more concerned with finding and exploiting vacant or undersubscribed economic niches in Britain as he had done in Africa. If this necessitated interaction with the indigenous population then this occurred, as did the adoption of more Westernized traits and outlooks when this was advantageous. The East African must not therefore be thought of as an embattled transient fighting to defend each and every element of his culture and identity.

CHAPTER 6

A Synthesizing Model

The previous two chapters have been concerned with different approaches to the explanation of South Asian residential patterns. One stressed that society is structured in such a way that access to scarce resources, such as good housing or desirable residential neighbourhoods, is restricted to those individuals whom society judges to be conforming and desirable. Conversely, those individuals who are deemed to be undesirable or even threatening are excluded, and forced to accept residual resources and life-chances. Coloured people in Britain fall into the latter category because they are not only from low-status families who are more often unemployed and who have more children, but they also suffer a colonial stigma. Consequently their options are restricted by society to the point where they are effectively compelled to live in the deprived and decaying inner city. The second approach contrasts markedly with the first. It takes a more anthropological line and asserts that the causes of contemporary patterns lie in the culture of the migrants, their aspirations, desires, and priorities. It argues that South Asian migrants came to Britain for temporary economic relief and that this demanded cultural isolation and frugality. As a consequence, their lives and mentality remained relatively unaffected by British society, the British class system, and racism. Many did not even realize the strength of feeling against them.

Chapters 4 and 5 demonstrate what these two approaches have to offer the analyst of Asian migration to Britain when they are taken in isolation. Clearly this is a good deal, but, having said that, they both seem to leave important questions unanswered when taken as separate entities. Why do ethnic divisions appear within classes whose members share similar access to social, political, and economic resources? And why do some groups retain or even strengthen their cultural identity when they find themselves in the immigrant situation whilst others happily relinquish theirs? These questions are not easily answered by either of these approaches as they stand, but are more readily understood when a combination of both sets of factors is taken into account. It is towards a *rapprochement* between these two lines of thinking that this chapter aims. However, before attempting this, there are a number of concepts and stances which must be described and discussed. Later in the chapter these will be used to modify and expand the two main approaches to Asian settlement as they are integrated into an holistic model.

Class Theory: A Constructive Critique

Perhaps one of the leading recent critics of the Marxist view of social differentiation is Parkin (1979). However, by suggesting what he considers to be a viable alternative to Marxist class theory Parkin avoids becoming nihilistic. He begins by outlining what he thinks is the prime objective of any theory of class: 'to identify the principal line of social cleavage within a given system—the structural "fault" running through society to which the most serious disturbances on the political landscape are thought to be ultimately traceable' (Parkin, 1979: 3). For Marx, this structural fault proceeded from the mode of economic production; those hidden forces which categorized individuals as either workers or exploiters. In seeing classes as the outcome of overriding economic forces, Marx explicitly downgraded those aspects of class and collective action which are attributable to the social characteristics of the individuals who are incumbents of the system. In short the system is propelled by its own logic and will continue regardless of the characteristics of those individuals who inhabit it. However, 'the awkwardness of this theoretical stance becomes evident whenever social groups act in blatant nonconformity with their assigned place in the formal scheme of things' (Parkin, 1979: 4). Here Parkin is thinking of racial, ethnic, religious, and sexual characteristics and how these are often seen to transcend objectively-defined class divisions. For Parkin, any model of class or stratification that fails to take these criteria into account loses all credibility since it contains a central hypocrisy: it is keen to perceive *inter*-class relations in terms of antagonism, dichotomy, and conflict yet it refuses to acknowledge that the same features might characterize *intra*-class relations. The latter must be seen as a bland concern with the niceties of social differentiation in 'a territory in which a truce has been declared in the *omnium bellum contra omnes*' (Parkin, 1979: 29). On the basis of this, Parkin (1979: 36) concludes that 'on current evidence it would be quite forgivable to conclude that the preferred Marxist response to the fact of racial or communal strife is to ignore it'.

Having rejected a Marxist approach to social differentiation on these grounds, Parkin suggests a reliance upon neo-Weberian notions, and the definition of classes in terms of their members' access to life-chances. Parkin, however, concentrates upon the idea of social closure and builds a class theory around this, which would account not only for inter- but intra-class conflict. Parkin describes the two forms of social closure which either a group or individual could adopt; these are usurpationary closure and exclusionary closure, the aim of both strategies being to restrict access to resources and opportunities to a limited circle of eligibles. Usurpationary closure can be defined as follows:

that type of social closure mounted by a group in response to its outsider status and the collective experiences of exclusion. What usurpationary actions have in common is the aim of biting into the resources and benefits accruing to dominant groups in society. What is entailed in all such cases is the mobilization of power by one group or collectivity against another that stands in a relationship of dominance to it. (Parkin, 1979: 74.)

Parkin cites as examples of this the collective efforts of ethnic groups to seek full inclusion into society, and the actions of an exploited proletariat to gain a greater percentage of the surplus value created by their labour. The latter are able to pursue such a strategy more effectively than the former for they are able to withdraw their labour selectively and therefore reduce the income of the dominant class. In contrast, the ethnic group frequently has less leverage since its members often occupy low-paid jobs which are weak in bargaining power. In addition, a substantial number may be unemployed and therefore unable to support collective action in the labour market. Ethnic groups are consequently forced to mobilize 'proxy power' by gaining some support from the dominant group through social and expressive mobilization. To gain success, such a strategy does presuppose the existence of a common political identity and collective consciousness within the ethnic group.

The second form of closure is exclusionary, and is defined by Parkin (1979: 45) as 'the attempt by one group to secure for itself a privileged position at the expense of some other group through a process of subordination'. That is to say, it is a form of collective social action which, intentionally or otherwise, gives rise to a social category of ineligibles or outsiders. Different criteria may be used to define this outsider group or, alternatively, exclusionary closure may be directed towards individuals. Common criteria at either a group or an individual level may be devices such as the ownership of property or the possession of qualifications or credentials. It is easier for a group to use exclusionary devices where it can define the target group as being alien and therefore outside the moral compass of the dominant class and it is also easier if the target group can be identified by visual or social cues.

Parkin argues that a description of a group's position should not be based upon its location in the division of labour but upon the character of its primary mode of social closure. A group is thus defined by whether its actions are predominantly exclusionary or usurpationary and by the target groups against which sanctions are directed. Defining groups in this way overcomes the hypocrisy of seeing inter- and intra-class conflict as being essentially different phenomena, since one group in a class may use dual forms of closure; exclusionary versus other fellow class members and usurpationary versus the dominant class. The definition of the group's status then depends upon which of these strategies is pursued more vigorously—if it is the former then the group must itself form a collectivity, whereas if it is the latter, individuals are clearly members of a group which is only part of a larger class.

Where the potential exists for dual closure, Parkin suggests that the form of closure which becomes dominant will be determined by relatively simple social and economic trade-offs. An example of this might be, 'Do I gain more from excluding other groups or by joining with them in usurpationary action against a third dominant group?'.

Parkin parts company with Weber on one point though. This is the criterion for closure. Weber argues that this can be any characteristic, but Parkin disagrees. He sees a legal definition of a group as inferior as being an essential first step towards making it an outsider group. Exclusionary closure is thus usually employed against groups who enjoy only marginal political status and those whose organizing and defensive power is weak. The making of an outsider group therefore involves the active participation of the subordinate group and the tacit legal support of the dominant stratum.

The notion of closure will be reintroduced later in the chapter, but in the meantime it is important to note that theories of class are not amenable to simple empirical testing and that they are subjective. Social closure seems a particularly relevant way of explaining a multi-racial society with inter- and intra-class conflicts, but it is by no means the only way of explaining such relations.

Ethnicity: Some Perspectives and Concepts

The 1970s saw a rebirth of interest in ethnicity as a concept and as a major, active, and permanent element in social and political differentiation. Until that period, what has since become known as the 'primordialist' school had held sway. In brief, this argued that the divisions within mankind were in many ways inherent and were the product of historical rather than contemporary forces—Milton Gordon's (1964) book on assimilation exemplified this stance well. In that book, he wrote:

> my essential thesis here is that the sense of ethnicity has proved to be hardy. As though with a wily cunning of its own, as though there were some essential element in man's nature that demanded it—something that compelled him to merge his lonely individual identity in some ancestral group of fellows smaller by far than the whole human race, smaller often than the nation—the sense of ethnic belonging has survived. It has survived in various forms and with various names, but it has not perished and twentieth century urban man is closer to his stone age ancestors than he knows. (Gordon, 1964: 24.)

However, the major problem for the primordialists was explaining how ethnicity *had* survived in the face of the forces of modernization which include such potent elements as migration, mass communication, mass transportation, and transnationalization (Rogerson, 1980). In short, what is the mechanism by which ethnic sentiments are communicated between generations and how are those most at risk prevented from abandoning their ethnic allegiances?

One piece of work in particular addressed itself to these questions and provided convincing and detailed conclusions: that is Mayer's (1961) study of Red and School Xhosa in the Locations of East London, Cape Province, South Africa. Mayer noted how the Xhosa were rural peoples who were forced to send migrants to the towns to supplement their meagre incomes. Whilst many of these male migrants stood by their indigenous way of life and pagan religions, others were more eager to adopt the ways of the white man and become literate and educated. The former were known as the *amaqaba* or Reds and the latter as the *abantu basesikolweni* or School people. Mayer was interested in how and why one group maintained its conservative outlook whilst the other rejected it, although both groups worked in similar jobs in the white economy and lived in similar areas of the city. Mayer began by investigating conditions within the sending villages and considering how society there reproduced itself. He noted that the composition of primary group contacts was pre-ordained and directly predictable from the structure of the village. Contacts were therefore largely with kin, neighbours, and age-mates and could not normally be contracted out of, unless by death. As Mayer (1961: 13) puts it, 'variable factors, such as accident or personal choice—variable in the sense that they and their effects cannot be inferred from knowledge of the social structure—therefore seem to be left with little to do for the *composition* of the network, within the rural community. All they do there is to influence the tone or emotional content of each of the relations in the network.' Moreover, in the village, institutions are undiversified, thereby restricting both choice, and the ability to opt out. With a limited and ascribed set of primary group contacts, and the absence of institutional duplication, relationships between individuals become multiplex in nature, and social networks are therefore close- rather than loose-knit (see Mitchell, 1969 for further explanation). Participation in close-knit networks exerts a consistent moral pressure on each and every individual to maintain a conformist 'tribal' morality. This in turn encourages only the development of undiversified institutions. The cycle of pressures is thus complete, and traditional ways of life and standards are maintained and transmitted to the younger members of society.

Mayer contrasted this situation with the one which prevailed when migrants left the village and moved to the towns for work. Here they came into contact not only with whites but with migrants from other villages who were also economically motivated and transient. For the first time, chance and choice could determine the individual's network, for this was no longer structurally predetermined as it had been in the village. As Mayer (1961: 14) puts it, 'now as never before the migrant chooses all the individuals who will be in *closest* personal relation to himself. If chance gives him his workmates, choice decides with which of them he will spend the lunch-break eating and chatting, or the evening visiting and drinking.' For the first time, physical proximity does not compel individuals to enter into social relations.

The same question of choice also arises with regard to institutions, which are more diversified than those of the village. The Xhosa must reconstruct a new network of relations, a new set of habits, and a new series of institutional memberships when he arrives in town. Yet he does all this, from the many possible alternatives with which he is confronted, in the absence of the direct moral pressure which has hitherto guided him in all such decisions. Whether a migrant chooses to abandon old ways depends partly upon his aspirations and his satisfaction with these ways. However, the thrust of Mayer's argument was that many Red Xhosa remain so 'incapsulated' during their residence in the towns that they either do not perceive School life as a viable alternative to their own, or the indirect operation of internal and external moral pressures prevents the conversion of potential choices into actual choices. Mayer went on to describe what he considered to be the key mechanisms of encapsulation: these were a negative evaluation of town life and a positive evaluation of village life; filial submission to parental wishes; the extended family; arranged marriages; and fear of the ancestors. Each of these factors ensured that when an individual left the village for town he carried with him internal moral sanctions which predisposed him towards conformity in the same way that the physical sanctions of the village had done. As Mayer (1961: 94) put it:

The special flavour of the incapsulated Red life in East London is due to the fact that it is viewed as a sort of continuation and extension of the home community life. The triumph of incapsulation is that through its institutions the home agents of morality have been enabled to extend their grasp over distance and over time, from the rural homestead into the heart of the East London slums. The long arms of the parents, the long arms of the ancestors, are constantly pulling the Red migrant back out of reach of the 'perils' of urbanisation.

However these internal moral predispositions are insufficient to guarantee that all Reds will make the 'correct' choices when faced with the multiplicity of alternatives within the town. A predisposition towards encapsulation is therefore strengthened by other mechanisms which convert this into reality.

One can discuss the mechanisms which serve to maintain incapsulation under two heads. The first is the insistence on actual contact with home by means of regular visiting. The second is the insistence on the solidarity of the 'people of one home place', the *amakhaya*, during their stay in town. These two practices interconnect—the *amakhaya* in town reinforce one another's resolve to keep on visiting home. (Mayer, 1961: 94.)

The behavioural concomitants of encapsulation, which simultaneously support it, are thus continual visits home at weekends, the sending of remittances, leaving children and wives in the village, residential clustering and room-sharing, group participation in leisure pursuits, the avoidance of contact with non-*amakhaya*, and the emphasis on a myth of return. In other words, the Reds maintain their ethnicity by consciously minimizing interaction with non-group members in almost every sphere of life. Spatial distance has an important role to play in this process, for it both creates and emphasizes social distance: Mayer (1961: 101) himself stressed this point when he wrote:

with Red people, the main determinant in all choices of room and room-mate, is the desire to stay close to the home-people. In the first phase satisfaction is gained from 'living with' one's *amakhaya*, in the second phase 'living near' them. Each group of room-mates therefore constitute a little cell of *amakhaya*. From these cells . . . neighbourhood clusters are formed, and also kinds of associative grouping which provide the framework for the organized life of Red migrants in town.

When drawing up his conclusions Mayer was not blind to the presence of structural constraints and their potential impact upon ethnic identity. However, he argued that these constraints were permeable and had been circumvented by many who had wished to do so. The central point was that the majority did not choose to do so, but chose instead to remain encapsulated within their traditional ethnic ways. As Mayer (1961: 286) concluded, 'the suggestion is, then, that the general relegation of Red migrants to the bottom of the class ladder in town does not *only* reflect their inability to rise by reason of lack of skill, but is also due to their persistence in acting-out parts according to the expectations of the home peasant society instead of the expectations of local . . . society'.

Whilst the primordialist's view was perhaps the most heavily subscribed to in the 1960s, different interpretations of ethnicity did exist and they did gain adherents. The 'circumstantialists' were possibly the largest alternative school.

They were doubtful that basic divisions did exist within mankind and looked instead to specific and immediate circumstances to explain why some groups maintained their identity whilst others did not. One line of thinking in this genre which is of particular relevance to our purpose here concerns the notion of 'marginality'. The concept was first introduced to the social-science literature by Robert Park (1928). He explained the circumstances which generated marginality as follows (p. 887): 'migration as a social phenomenon must be studied not merely in its grosser effects, as manifested in changes in custom and in the mores, but it may be envisaged in its subjective aspects as manifested in the changed type of personality which it produces'. He described (p. 892) this personality type as

a cultural hybrid, a man living and sharing intimately in the cultural life and traditions of two distinct peoples; never quite willing to break, even if he were permitted to do so, with his past and his traditions, and not quite accepted, because of racial prejudice, in the new society in which he now sought to find a place. He was a man on the margin of two cultures and two societies, which never completely interpenetrated and fused.

Park's idea that contact between two cultures would produce conflict which might then be mirrored within the individual was later followed up and expanded by Stonequist (1935). He described common marginal-personality traits as increased sensitiveness, self-consciousness, race-consciousness, *malaise*, and inferiority. Moreover he went on to describe a cycle through which the marginal man was thought to pass: this began when the individual was introduced to the new culture and unwittingly began to be assimilated into it. In time, though, a 'crisis' would make him aware of cultural conflict (either through one single experience or through a gradual process of realization). This would be followed by feelings of shock and confusion, and an attempt by the individual to readjust his personality over the long-term. The strategies which presented themselves at this point were complete assimilation (made easier if the individual lacked biological markers), 'passing', a wholehearted return to the original culture (with a residue of resentment), or complete withdrawal with other like-minded individuals into an isolated and self-contained community.

Of greater consequence for the sociological, as opposed to psychological, study of migrants, was Goldberg's (1941) paper on the marginal man. He separated out the two elements of the marginal-man concept and considered them as separate entities which were not necessarily linked, as Park and Stonequist had implied. In particular, he argued that what might be thought of as a marginal *position* by sociologists need not necessarily produce the marginal *personality traits* which Stonequist had elaborated. If certain conditions were met, a marginal situation could produce a satisfying and 'normal' culture.

The exact relationship between personality and sociological marginality continued to be the subject of much debate during the ensuing decade. Green (1947) considered that the personality was only affected when an individual came up against barriers to his entry to the superordinate group. Kerckhoff and McCormick (1955) pointed to the importance of both individual orientation and the attitude of the dominant group. Child (1943) concurred. Mann (1957) concluded that marginal personality traits could result from group relations but also from other causes. Antonovsky (1956) showed that different marginal positions produced a range of responses, some of which involved the personality whilst others did not.

Simultaneously with this expanding knowlege of the link between sociological and psychological marginality, the term was also re-defined to include groups other than migrants. Hughes (1949) for instance included Bohemians, parvenus, black academics, and unmarried career women. Others thought such definitions too broad (Golovensky, 1952), and Braithwaite (1960) even went so far as to suggest that the concept had lost all utility.

Despite the continuing dialogue about the psychological importance of marginality, the nature of the marginal situation and its effects on group structure and function remained largely unresearched until the publication of Dickie-Clarke's (1966) study of Durban Coloureds. Dickie-Clarke sought to establish a purely sociological definition of the situation divorced from the 'misleading stereotype of the racially—or culturally—mixed ''marginal man'' as a kind of unhappy, half-pathetic, half-comical caricature of the ideal to which he is presumed to aspire' (Dickie-Clarke, 1966: 22). His first step was to distinguish between the different elements contained within the notion. This he did by suggesting three definitions, the first of which concerned the structural position which could be termed marginal: this

refers to the fairly long-lasting, large-scale, hierarchical situation in which two or more whole groups or even nations exist together. The groups vary in degree of privilege and power and there is inequality of status and opportunity. The barriers between the groups are sufficient to prevent the enjoyment by the subordinate group, or groups, of the privileges of the dominant, non-marginal group, but do not prevent the absorption by the former of the latter's culture. (Dickie-Clarke, 1966: 21.)

The second meaning which Dickie-Clarke ascribed to marginality referred to what this larger group situation entails for the individual. Is one part of a community of people, or an isolated individual? Are barriers to mobility legal or customary? Is the marginal situation temporary or relatively permanent? Thirdly, marginality could be defined as the way in which an individual reacted to being in a marginal situation. For some this would entail personality traits or disorders whereas for others it would not.

Dickie-Clarke then went on to discuss only the first of these three possible definitions and sought to relate it to general stratification theory. He argued that the marginal situation was a special case of hierarchical situations, but differentiated from them by the presence of inconsistencies in the ranking of the marginal individuals or groups. These inconsistencies might be either cultural or social, and could be ascribed or achievable. If they were ascribed then the marginal situation would be long-lasting, if achievable then it would be short-lived and of lesser importance. If inconsistencies were confined to different rankings of a collectivity according to various criteria *within* either the cultural or social category then lesser forms of marginality would arise. If, on the other hand, inconsistencies were between how a group was ranked socially and culturally then the resulting marginality would be gross. Finally, Dickie-Clarke thought that marginal situations were likely to be more complicated than a simple structure which involved only two groups, one of which was superordinate. Moreover,

simply because of the very dominance of the superordinate stratum, those marginal situations in which the inconsistency is with the dominant stratum are more significant and important than those where the inconsistency is merely between two subordinate strata. It is the dominant stratum which counts in most matters and usually the hierarchy is designed to protect its dominance. (Dickie-Clarke, 1966: 43.)

Having laid down rules which categorized significant and less significant forms of marginality, Dickie-Clarke applied these to the Durban Coloureds in an effort to show how the concept of marginality might elucidate a group's structure and social position. He noted that the Coloureds were only one of three subordinate groups in South Africa (the others being the Indians and Africans), that the hierarchy embraced the whole of society and all aspects of life within it, and that the barrier between whites and Coloureds had allowed the latter to become cultural but not social equals. He described their position as being a marginal one in a double sense

for they are in a marginal situation because there is inconsistency between their cultural equality and their social inequality and furthermore, if the social dimension is taken by itself, there is inconsistency between their ranking of equality in some matters (e.g. certain civil rights) and their lower, unequal ranking in other matters (e.g. pay and employment opportunities) (Dickie-Clarke, 1966: 105).

Dickie-Clarke then looked at the way in which their position as a marginal group affected the Coloureds' behaviour. He suggested that the inconsistencies produced feelings of rejection and victimization. In particular, the Coloureds have developed similar attitudes towards Indians and Africans to those held by the whites. They thus seek social distancing and residential segregation from them but the Coloureds' economic position in society does not allow this. The white superordinate group thus forces the Coloureds to reside and interact with groups whom they have taught the Coloureds to regard as backward. It seems natural that insecurity of status develops as a result.

The end product of a marginal situation for ethnic sentiments varies a good deal according to the criteria adopted for acceptance, the degree and form of inconsistencies, and the pre-existing attitudes of the subordinate group. Some groups might withdraw into a more encapsulated position and emphasize the defence of their ethnic identity. Others might react by emphasizing their ethnicity in an assertive or aggressive manner, in an attempt to gain entry as equals. The increase or diminution of ethnic sentiments can therefore be seen as a direct response to the circumstances prevailing at the time.

The second major school of thought to challenge the primordialist stance on ethnicity has yet to be named but centres upon the ideas of a 'new ethnicity' as proposed by Glazer and Moynihan in 1975 and subsequently expanded by other authors in the same collection of papers. At its simplest, Glazer and

Moynihan's argument is that ethnic groups are not mere survivals from an earlier era but a major new element of social differentiation which is likely to usurp the position of class and status as the pre-eminent criterion for defining interest groups. According to Glazer and Moynihan, this change had occurred because of the increasing inadequacy of existing interest groups. Classes or statuses were frequently too large and too ill-defined to generate precise consensual claims on society's resources. Moreover, if such claims were met, the benefits were diffused amongst too many claimants. In addition, ethnic mobilization had a significant advantage over other forms of organization in that it offered a combination of interest with an affective tie. Again this ensured that ethnic groupings were a more efficacious way of pressing demands. For Glazer and Moynihan then, ethnicity was not a matter of tradition or sentiment but a strategy which offered a more effective way of claiming, gaining, and distributing society's resources.

An Holistic Model

Having described the two opposing approaches to Asian settlement in Britain, and discussed a series of concepts and ideas which seem to have been excluded from extant studies despite their apparent relevance and value, it remains to create an holistic model which aims to reconcile the two approaches whilst retaining their obvious advantages and insights. Figures 6.1 and 6.2 illustrate such models and each of the subsequent sections describes some major element of them.

British society prior to the arrival of coloured immigrants

Figure 6.1 demonstrates the way in which British society was (and is) differentiated by both class and status. The former is a discrete phenomenon whilst the latter is continuous in nature. The middle class, which can crudely be divided into the 'comfortable' (such as professionals) and the 'striving' (low-paid clerks), believes in allocative criteria centred upon notions of merit or achievement. The mechanism which they see as the arbitrator in conflicts over scarce resources is guided free enterprise, and the institutions and individuals which they feel embody this spirit are exemplified by the Tory party and societal gatekeepers. The latter are often professionals recruited specifically to discriminate between 'deserving' and 'undeserving' candidates. The criteria for access to the middle class are birth (through inheritance of family influence), education, and enterprise, although the last of these may be replaced by perceived social worth, especially in occupational ranking. The preferred or respected route to mobility both into the class, and within it, is through individual action. The dominant form of social closure in which the class participates is exclusionary, and is directed at a defence of inherited privileges. In the sphere of housing this involves protecting the social status of neighbourhoods of owner-occupied properties and exclusive privately rented accommodation. This is accomplished through the price mechanism (or 'free market forces') and the aid of urban gatekeepers, such as estate agents and building society staff, who ensure that 'conforming' applicants are placed in 'suitable' areas.

The British working class differs significantly from the middle class. It believes in the allocative criterion of need, and an allocative mechanism of state intervention in the form of a planned welfare system. The institutions which it feels supports these ideals are the Labour party and the Trade Union movement. The criteria for access to the working class are birth (through the cycle of disadvantage), default, idealism, and downward social mobility. The preferred form of staking a claim for greater resources is through collective mobilization, exemplified in the economic sphere by the withdrawal of labour. The predominant mode of social closure is usurpationary; in terms of the housing system, this means claims for a shift in balance towards public ownership of

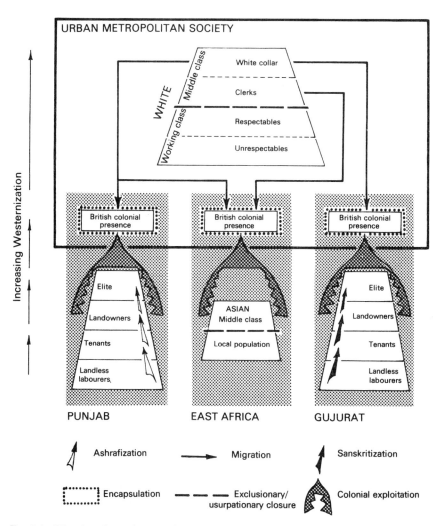

FIG 6.1: The situation prior to Asian migration to Britain

housing, for reduced council rents, for control of private rents, and for extension of subsidized schemes such as local authority mortgages and central government improvement grants. These claims for greater resources are likely to be pursued at a local and national level by formal groups, bodies, or political parties, rather than by individuals. Again, in certain circumstances, the 'respectable' working class rely on professional gatekeepers to ensure that 'deserving' applicants who are 'in need' are given suitable resources.

Because of the demands of Empire, members of both classes, but more especially of the middle class, had seen service overseas either as teachers, administrators, or members of the armed forces. There they had come into limited contact with the local population and had developed what were often ill-informed opinions about them. Moreover they had contact with the local population only within the context of colonialism and they therefore viewed them as natural inferiors and occupiers of roles which enjoyed only a low status (see chapter 4). Finally, contact was frequently the result of military victory and the imposition, by force, of an alien regime. At best this resulted in a benignly paternalistic attitude, and at worst, callous arrogance and disregard (see Blauner, 1969 on the link between colonization and racism). The urban metropolitan society, of which the middle and working classes were a part, thus internalized attitudes towards Indians and Pakistanis which were shaped by colonial exploitation and expressed in derogatory stereotypes. This is not to say that all individuals held such beliefs and attitudes but without doubt these were the prevailing ones which guided the society's behaviour.

South Asians and society during colonial rule

Punjabi and Gujurati society is conceptualized as consisting of five different statuses, one of which was implanted (see figure 6.1). The encapsulated British colonial representatives remained aloof from the indigenous peoples and practised social avoidance and social distancing. They may be thought of as a satellite of British middle-class society serving out a period of overseas residence before their eventual return to England. Because of this belief in a myth of return they made few social or cultural concessions to life in India and East Africa. Within the sub-continent the remainder of the population has been divided into four statuses dependent upon their wealth and social influence. The local élite in both societies is the group which has most contact with the British representatives and it is also the group which contains the highest percentage of individuals aspiring to Westernization. As was outlined in chapter 5, this involves an emulative life-style based upon wealth and formal education. The other three statuses also aspired to such a life-style in varying degrees, dependent upon the extent of their contact with British representatives and the strength of their imitative tendencies. In general, though, these groups had more immediate priorities which were internal to traditional Indian society prior to Empire. In Gujarat these priorities would be the acquisition of wealth and land to allow ritual mobility, or Sanskritization, whilst in Punjab the same wealth and land would be used to achieve Ashrafization. Coupled with this, both the landowner and tenant would be anxious not to become *kami*, or landless labourers, whilst the entire society was orientated towards achievement and upward mobility. One strategy to achieve this was voluntary migration to the British East African colonies either to gain a new life there, or to earn money prior to an eventual return home.

As chapter 3 pointed out, the circumstances of Asian settlement in East Africa were at least materially, if not socially, superior to conditions in the sending society. The group was regarded as akin to a middle class and had accumulated considerable wealth through middle-man entrepreneurial activity. They had consorted with the British representatives to make colonialism successful and profitable, and shared many common views with the British about the local population, which became the subject of stereotyping. Although the East African Asians aspired to a more Westernized life-style than their sub-continental counterparts, they were still not treated as equals by the local Britons who maintained largely economic contacts with them and indulged in derogatory stereotyping of them. Lastly, towards the close of colonial rule in East Africa, the rise of African nationalism set the local and Asian populations on conflicting routes. The former adopted tactics of usurpationary closure (such as nationalization) while the latter pursued exclusionary closure to prevent the erosion of their privileged position.

Post-colonial Asian society

The removal of British representatives from Gujurati and Punjabi society did little to upset its structure or the aims of its members. One might conjecture that Westernization filtered a little further down the hierarchy and displaced Sanskritization and Ashrafization, but this cannot be proven. What can be shown though is that economic, and therefore social, pressures became intensified as a result of the burgeoning population. The search for strategies to avoid loss of *izzet* took on a new significance and introduced large-scale voluntary target migration either to urban centres or to Britain. Chapter 5 discussed how these migrants perceived British society, and their attitudes towards permanent settlement there. In this context, the myth of return is of central importance.

As was pointed out in chapter 3, the situation in East Africa was very different and ultimately produced an involuntary exodus of Asians, many of whom came to Britain.

British society after the arrival of coloured immigrants

The arrival of Asian immigrants in Britain as a direct response to selective labour shortages, served only to emphasize and strengthen pre-existing prejudices. Whilst the white middle class continued to focus upon exclusionary closure towards the working class, it did provide a lead to the latter on the issues of attitudes to coloured workers. The early, and open, discussion of restricting immigration, and the failure of successive governments to grapple with the problem of discrimination served to legitimize the differential treatment of

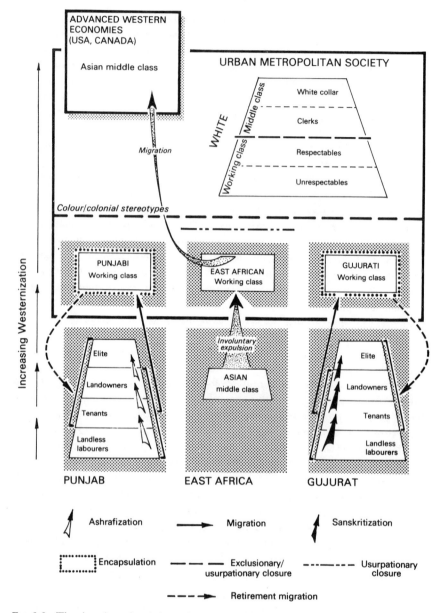

FIG 6.2: The situation after Asian migration to Britain

coloured people. This was emphasized by the way in which coloured immigration and immigrants were seen as a 'problem' and also as a way of gaining political advantage. Although there were undoubtedly some liberals within the ranks of the middle class, the majority continued the pre-existing colonial relationship with Asians, demanding cheap labour and offering little in return in the way of social acceptance or resources.

The problem for working-class whites was on a different plane. Whilst they had continually been told that Asians were their inferiors, some were now forced into a position where they had to share accommodation, neighbourhoods, occupations, schools, and associations with members of that group. They were thus confronted with the fact that their position in life differed little from that of a group they had always regarded as natural inferiors. This must inevitably have undermined confidence and self-identity. In addition, though, an increasing number of the indigenous working class were thrust into situations where they were in a perceived conflict with the immigrants for jobs, houses, and other scarce resources (see chapter 4). The white working class thus had to endure simultaneous attacks on its identity, self-perceived status, and material position. The net result of this was that the class entered into a position of dual closure where its pre-eminent actions were aimed at gaining a greater share of resources at the expense of the middle class, whilst it also engaged in exclusionary closure towards coloured immigrants. Exclusion took the form of prejudice, discrimination, calls for a reduction in immigration, changes in voting behaviour, and social avoidance. Pre-existing colonial attitudes would encourage this and stereotyping would legitimize it. Colour provided a useful visual marker and social cue.

The background to the attitudes which Indian and Pakistani migrants brought with them to Britain has been referred to in chapter 5. Suffice it to say that they arrived with a negative evaluation of British society, a very positive evaluation of their own culture, a belief that their stay in Britain was only temporary, and an in-built tendency to encapsulation. In the light of this, it seems understandable that they remained largely unaware of the forces of exclusionary closure which barred their potential social and cultural mobility. Moreover their orientation towards the sending society and its criteria for mobility ensured that few engaged in usurpationary closure against the white working or middle classes: chapter 5 outlined how Asian politics in Britain were largely focused upon issues which were current in the sub-continent, and not upon procuring a greater share of available resources whilst in Britain. In addition, that chapter contained a discussion of how the belief in return migration engendered within the Asian population a desire to avoid sources of cultural threat, whether these be external or internal to the group. Indians and Pakistanis therefore engaged in what could be termed encapsulatory closure, designed to maintain or strengthen cultural boundaries and faiths.

Finally, East African Asians arrived in Britain without a myth of return and as settlers rather than target migrants. Moreover the conditions from which they had come produced higher expectations of life in Britain and greater demands for social and financial mobility. Finally, they had left a society where they had been treated as somewhat superior by the British administrators, and had achieved middle-class status. Taken in conjunction, this would stimulate a policy of exclusionary closure towards Indians and Pakistanis who might be thought of as uneducated peasants. The creation of social distance would act as a way of avoiding white stereotypes of Indians and Pakistanis, and therefore of being 'tarred with the same brush'. Simultaneously, the East African Asians are engaged in usurpationary closure towards the white working class, as they attempt to regain a more affluent life-style and its associated social prestige. In the short-term, though, the East Africans are confronted with significant status inconsistencies: their internalized view of their own social status no longer matches up to that of the wider society; their cultural equality is not paralleled by social equality and there are also inconsistencies within the degree of their social acceptance. Whilst society recognizes them as social equals in some domains, it clearly does not in others. The East African Asian is aware of these contradictions, inconsistencies, and barriers, and therefore of his marginal position in society. This, and restricted economic opportunities in Britain, have forced many East African Asians to move on from here to other advanced western economies, most notably Canada and the USA.

Finally, for the sake of completeness one should mention the position of first and second generation West Indians in British society. Here again, significant differences appear to divide the population. Many of the first generation still argue that usurpationary closure through the institutions of the wider society must be the route to social advancement. Others (and more particularly the black youth) argue that usurpationary closure must take non-political and more aggressive forms if it is to succeed. This is one possible explanation for the inner-city riots of 1981 (see Taylor, 1981; Blauner, 1969). Finally, others take the line that white society is unlikely ever to accept them and therefore they withdraw into alternative cultures such as Rastafarianism (see Cashmore, 1979; 1981) which are characterized by encapsulatory closure. Pryce (1979) has provided the most detailed description of these varying reactions within the West Indian population to marginality and exclusionary closure. He identified six differing life-styles which could result, and grouped these into two major orientations: the 'stable law-abiding orientation' and the 'expressive-disreputable orientation'.

Some hypotheses

Having described the characteristics of an abstract model designed to elucidate the position of Asian immigrants in British society, it remains to map into this a segment of the real world via case-study material. Recourse to empirical data should emphasize not only that the model fulfils Schutz's postulate of adequacy by being grounded in, and derived from, real life, but also that it provides a useful function in aiding the visualization and explanation of complex relationships and interactions. These are, of course, two of the main purposes of a model, and two of the major criteria on which it should be judged. However, it is not possible within the confines of a work of this nature to re-analyse all aspects of the position of Asian immigrants in British society in the light of the model proposed above. Instead, certain key areas will be isolated and tested against empirical data, whilst others which have already been described and tested at length elsewhere, will have to be taken as given. In these cases, it is hoped that chapters 4 and 5, and the literature cited in them, will have provided convincing arguments for acceptance. Topics within this category, which fall outside the scope of this book, are the exact form and detail of the relationships between the white working and middle classes, the specific content and occurrence of Sanskritization, Ashrafization, and Westernization, and further analysis of the development of stereotypes during the colonial era.

In short, testing of the model will be confined to those issues which directly impinge on the every-day life of Asians in urban Britain. Other issues will be touched upon and referred to where necessary, but in less detail. Moreover, since it is clearly impossible to map into the model the whole of British society, case-study material will be drawn from one settlement only, namely Blackburn. This case-study material will be used to investigate a number of specific issues raised by the model.

Firstly, whether there are significantly different reasons for different sections of the Asian population being in Britain, and whether this influences their orientation towards both the host society and the remainder of the Asian population. In other words whether it is more appropriate to categorize South Asians and East African Asians as transients, settlers or refugees.

Secondly, whether structural and cultural forces combine in Britain to produce a situation where all Asians may objectively be regarded as a deprived underclass which is excluded from effective and equal participation in society and the distribution of scarce resources. Further, whether this exclusion may reinforce pre-existing cultural tendencies towards non-participation, and how these barriers may, to a certain extent, be tangential since they exclude some members of the Asian population from resources for which they have little desire. In this sense, structural and cultural forces interact to produce similar results.

Thirdly, whether because of contrasting attitudes, strategies, and behaviour South Asians might usefully be thought of as being encapsulated whilst East African Asians could more appropriately be linked to the concept of marginality.

Fourthly, whether these differences in attitudes and strategies are also found at a finer scale of disaggregation within the 'South Asian' and 'East African Asian' population and therefore whether an ethnic continuum could be developed along which the variety of regional-linguistic groups could be ranged.

CHAPTER 7

The Migration to Blackburn

The theoretical discussion of structural and cultural forces presented earlier in the book suggested that one cause of the lack of Asian competition for scarce resources is that the migrants continue to retain their own cultural values and preferences after their arrival in Britain. Retention of cultural values for a limited period after migration is not unusual and most models of assimilation assume such a phase (see Price, 1969). However the model put forward in chapter 6 assumes not a gradual transition towards assimilation, but the retention by many Asians of the standards and norms of the sub-continent for considerably longer than might ordinarily be expected. The reasons for this apparent anomaly include the motives behind migration and the perceived permanence of residence in Britain. It is these issues which are the central concern of this chapter. However whilst it is clear that there are significant differences between the values of the white and 'Asian' population in Blackburn it must also be made clear that the notion of a unified and homogenous 'Asian community' exists only in the minds of whites. 'The' migration of Asians to Blackburn is thus a false concept. There are in fact a number of different migration streams, members of each of which can draw upon varying pre-migration experience and differing aspirations. This chapter will try to draw out the contrasts in the purpose, permanence, and timing of 'South Asian' and 'East African Asian' migration which were enumerated and detailed in chapter 3. Chapter 9 will reveal whether these contrasts are reflected in residential and non-residential patterns.

Published national data are not suitable for such a task but they do indicate that in 1971 the County Borough of Blackburn contained some 5220 individuals of NCWP origin. Of these, approximately 82 per cent were of Indian or Pakistani descent. By 1981, these figures had risen to 16,039 individuals and 95 per cent Indian or Pakistani descent. However these data provide no insight into *why* South Asian and East African Asians came to Britain and to Blackburn. To answer questions of this nature, two special surveys of the Asian population in Blackburn were commissioned in 1977 and 1978 (see Appendix II for details). The former will henceforth be termed the 1977 Asian Census and the latter will be called the 1978 Sample Survey. These data, which relate to 8862 adults and children, and 391 households or part households, respectively, form the data base upon which the remainder of the book has been constructed.

Motives for Migration

Chapter 3 provided a detailed account of why South Asians *chose* to leave their homelands, and why East African Asians were *forced* to leave theirs. It outlined

116

the economic and status considerations which motivated Indian and Pakistani target migrants and it described the involuntary refugee status of East Africans from Kenya, Uganda, and Malawi. These general reasons are borne out in the 1978 Blackburn sample, and are tabulated in table 7.1.

As the table makes clear, the key reasons for East African Asian migration were political push factors. Thirty per cent of respondents named these as the prime motive and a further 12.5 per cent said that they had come to Britain to improve the quality of their lives. For many, this would be synonymous with an escape from racial harassment. Specific comments recorded in the 1978 Sample Survey included 'no future in Africa', 'unable to get a licence for my shop in Malawi', 'Government closed my business down', and 'too many restrictions for Asians in Africa'. These comments also hint at a second under-lying cause of the migration, the desire for economic and social advancement. Twenty-five per cent of respondents said they had come to Britain to develop their careers, and a further 10 per cent to take advantage of British educational facilities. Clearly, then, whilst many East African Asians were victims of circumstance, this should not be allowed to mask the competitive nature of the group as a whole.

In contrast, the South Asian migration stream seems to have been motivated to a much greater extent by financial needs and associated ambitions towards family advancement. Nearly 49 per cent of respondents had left the sub-continent either to find work or to find better paid work. A further 7.6 per cent admitted to coming to Britain for purely financial reasons, which included 'to make money', and 'to support my family back home'. Ten per cent came 'to improve their lives', a category which could include religious persecution as well as financial considerations. Lastly, nearly 7 per cent migrated to improve their education, a reason linked more to 'Westernization' than 'Ashrafization' or 'Sanskritization'. A second point worthy of note is that the migration of South Asians shows clear signs of its chain or circulatory nature and of the fact that it was a mass phenomenon; 21 per cent of respondents had come to join or replace relatives and friends, whilst three interviewees admitted that they had left India simply because everyone else in their village was doing so. Four people even said that they did not know why they had come to Britain.

The differing purpose of migration for the two groups is also made apparent by an analysis of their timetable of arrival in the UK. Table 7.2 provides a detailed breakdown of this information. It demonstrates that the first South Asian respondent to come to Britain had done so in 1951, but that the migra-tion had not really begun to gather pace until the late 1950s and early 1960s. It also shows that the 1962 immigration legislation had less impact than is widely thought. However the legislation later in the decade appears to have been considerably more effective, for it shifted the emphasis away from primary migrants (or the heads of household, from whom the Blackburn surveys collected data) towards dependants. As a result, relatively few of the South Asian sample arrived in the UK during the 1970s. Overall then, nearly 46 per cent of South Asian respondents had already migrated to the UK prior to the imposition of the 1962 controls, a further 39.3 per cent had arrived before the 1968 Act, and only 14.8 per cent had arrived since. This pattern is in line with national trends outlined in chapter 3. The East African sample demonstrates very different characteristics. The first respondents arrived as late as 1958 and were not followed by others until 1963. Consequently only 4 per cent of East African interviewees had entered Britain before the beginning of 1964. The migration only gathered momentum after 1967, when Kenyanization began to have a significant impact upon the lives of Asians in that country. This first wave of immigration was, however, rapidly ended by the 1968 Act and the imposition of quotas. Thirty-six per cent of respondents had succeeded in gaining entry to the UK between 1964 and 1968. A reduced number continued to gain entry despite the 1968 Act and this steady flow was augmented by panic migration in 1972 from Uganda, and 1976 from Malawi. The combination of these flows meant that 60 per cent of the East African sample had migrated to Britain since 1969. It seems then that, both in our sample and at a national level, East African immigration was linked to political push factors whilst the South Asians came to Britain in order to gain social and financial mobility during a period of economic opportunity.

In addition, the 1978 Sample Survey allows a more detailed consideration of the migration histories of those who came from East Africa. The country in which the largest number of East African respondents had lived, at some time or other, was Kenya (28 households), followed by Malawi (17), and Uganda (13). This again confirms that the migration from East Afica was largely motivated by programmes of Africanization.

Chapter 5 described how the motivation behind South Asian migration was the desire for social mobility and how migration was therefore seen as a temporary expedient to achieve specific aims. It also demonstrated how the target migrants organized their lives and thoughts on this basis and consequently had their behaviour controlled by a myth of return. Since they believed that they would ultimately return to their village as successful and respected *vilayati*, they would ensure that the patterns of behaviour and norms which had guided them prior to migration would continue to do so after migration. The 1978 Sample Survey contained a specific question about the initial intention to settle in Britain on arrival. However, the results are not easy to interpret. Approximately 82 per cent of both the South Asians and East Africans said that, when they first arrived in Britain, they intended to stay here. Fourteen per cent of the latter group had no intention of doing this, because they saw Britain as a staging post on the route to joining relatives in Canada, America, or Australia. Many indicated that they were actively engaged in making arrangements to accomplish this last stage of their international migration in the near future, although it is difficult to know whether the onset of the economic recession in Britain would hasten or postpone such moves. In contrast, those 16 per cent of the South Asian sample who were convinced that they would not stay in Britain, were predicting that their economic targets would be reached relatively rapidly and that they would therefore return shortly to the sub-continent. It is difficult to know whether the 82 per cent of South Asians who thought that they *would* stay in Britain envisaged permanent settlement there, or whether they were simply expressing the view that it would take the majority of their working lives to achieve their economic targets and therefore that return migration was an eventual but far too distant possibility. A clearer indication of intentions can, however, be gathered from a supplementary question which was also contained in the 1978 survey. This was the direct question, 'Do you still intend to return home?'. Again, no time period was specified. Almost 37 per cent of South Asians answered in the affirmative, whilst the comparable figure for East Africans was only 23 per cent. One can conclude, therefore, that the myth of return is a part of the value-system of nearly four out of every ten South Asians in Blackburn, but within the East African population, 'home' whether defined as Africa or the Indian sub-continent, seems to have a significantly weaker attraction.

The conditions under which migration took place, and attitudes towards its permanence, also influence the demographic structure of the two Asian populations. The economic nature of South Asian migration should have its corollaries in the sex and age structure of that migration stream. The average age at which respondents had first entered Britain was 24.3 years. The youngest of those who were now heads of household had left the sub-continent when he was six years old, and the oldest had been 65. The average age for migration from East Africa was noticeably higher at 32.5 years. The oldest head of household had been 73 years of age on arrival, and the youngest, ten. Only one South Asian had arrived in Britain as a retired dependant whereas 10.4 per cent of East African migrants fell into this category. These data again indicate that initially migration from South Asia was undertaken by those who were capable of earning a living, and therefore able to support themselves and their families at home. East African migration was of a different nature and, because of circumstances, had to be less selective. It also involved all members of the community, including dependants. However, as contemporary accounts illustrate, the responsibility for accepting the East Africans for settlement in Britain was neither immediately nor completely accepted by the government. Many were forced to settle in India and accept an undertaking that they would eventually be allowed into Britain when the size of the quota permitted this. The Immigration Section of the British High Commission in Delhi says that it is not allowed to reveal either the size of the quota allocation to East African Asians awaiting entry from India or the number of people who have applied for such entry. They *are* allowed to comment that those applicants who were being issued with entry certificates in February 1982 had first lodged applications in 1978. As a result, whilst migration from East Africa to Britain *was* less selective and more complete, there are still dependants awaiting entry, and the population in Blackburn is therefore not yet in demographic balance. Slightly more than 22 per cent of East Africans in Blackburn still have dependants seeking entry. The less balanced nature of South Asian migration, especially in its earlier phases, is reflected in the fact that nearly 30 per cent of such respondents were still awaiting the arrival of dependants. Thus whilst the migration of East African Asians to Britain was *en bloc*, that of South Asians was more determined by the requirements of the labour market.

Similar factors operate when looking at the age and sex structure of the two populations as they exist in Blackburn at the present moment. The 1977 Asian Census enumerated 577 adult males of East African ethnicity (as opposed to birth) and 496 adult females. There is consequently a sex ratio of one male to 0.86 females within this group. Comparable figures for the South Asian population were 1873 adult males and 1524 females, or a sex ratio of 1:0.81. This indicates that despite their earlier arrival in Britain, family reunion within the South Asian group has not yet been sufficiently widespread to produce a sex balance similar to that found amongst East Africans. The same can also be said of the age structures of the two populations. As table 7.3 illustrates, South Asian migrants have not, on the whole, been joined in Britain by their aged relatives. Only 1.3 per cent of South Asian adults are over 60 years of age, and the single most predominant characteristic of the group's age structure is its youthfulness. Almost 40 per cent are less than 26 years of age. The population of East African ethnicity also has a strong youthful nature but this is balanced by a larger component of more mature adults. Nearly 13 per cent are over 50 years of age and 6.5 per cent are more than 60 years old. The presence of a larger number of older East Africans is also reflected in the average household size of this group. The 1073 adult East Africans enumerated in the 1977 Asian Census lived in 333 separate households, the average size of which was consequently 3.22 adults. In comparison, 3397 adult South Asians lived in 1253 households, giving an average 2.71 adults per household. Amongst the East African population, the most frequent household type was the two-generation extended household containing head and spouse, plus the parents of either the head or his wife. The next most frequent was the nuclear household. Amongst the South Asian population, nuclear and two-generation extended households did exist but these were in addition to various one-generation extended households which contained combinations of brothers, sisters, cousins, and young in-laws.

The circumstances of East African migration to Britain also had a major influence upon conditions and life-styles immediately after arrival in this country. The fact that emigration had not been planned meant that employment and housing could not be pre-arranged, and although some of the early expellees were successful in their attempts to bring capital with them, many were not. Ninety per cent of the East Africans questioned in the 1978 Survey arrived in Britain without pre-arranged employment whilst a further 5.3 per cent were still engaged in full-time education and were not, therefore, seeking work. When they did find employment, the East Africans relied less upon community support or official methods (such as Job Centres and Training Centres) and more upon their own resources. Sixty-one per cent of respondents said they had found work on their own initiative and a further 25 per cent located work through advertisements or white contacts. A not dissimilar pattern emerges in the field of housing, although nearly 47 per cent of respondents had been forced to rely upon their relatives for temporary accommodation when they first arrived in Britain. Twenty-five per cent had sufficient resources to purchase a property immediately, but more than 14 per cent had been forced to find some form of privately-rented accommodation. This contrasts with only 5 per cent of the South Asian population. In the longer term, though, the East Africans again chose to make less use of community support during the process of finding permanent accommodation. Almost 60 per cent of respondents had found their own accommodation after arrival either by answering private advertisements, making enquiries in shops, or using the services of an estate agent.

The differing nature of the two migrations is also reflected in why the groups chose to settle in Blackburn rather than elsewhere in the country. Forty per cent of the 1978 South Asian respondents considered that their prime motive had been financial. Included in this category are factors such as 'more jobs available now', 'wages better here', 'low cost of living', and 'availability of cheap housing'. One respondent was even able to give a breakdown of the relative cost of property in a number of neighbouring towns and show that Blackburn was significantly cheaper. Fifty-three per cent cited the presence of a sizeable Asian community in Blackburn as the main feature which had attracted them to the town. The East African respondents gave a wider range of reasons for their decision but only 30 per cent mentioned financial motives. Over half mentioned the presence of friends and relations and the help which they would be able to afford in the early stages of settlement. Many now thought that they would move from Blackburn since their adjustment to life in Britain was sufficiently advanced to allow a greater measure of independence. Leicester seemed to be the favourite proposed destination because of its sizeable, even dominant, East African community and the less conservative atmosphere which this group generates. Leicester was even described, in glowing terms, by one respondent as 'Sin City'. Clearly, such attractions outweigh the financial incentives of life in Blackburn. During the early phases of settlement though, the East Africans, no doubt especially the expellees, were content to settle wherever they could. More than 10 per cent of respondents admitted that they had drifted to Blackburn without especially strong motives, or without any motives whatsoever. Less than 2 per cent of South Asians admitted similar reasoning, or lack of it.

Backgrounds and Potentialities

A key to the roles which the East African and South Asian groups are likely to play within the Asian population as a whole, is the background from which members have come, and the experience which they bring with them. The 1977 Asian Census enquired about whether migrants had left a rural or urban life-style to come to Britain. Seventy-six per cent of South Asians classified themselves as rural, and 24 per cent as urban. This underlines the essentially agrarian background from which Indian and Pakistani migrants are drawn. One should also bear in mind that the respondents themselves were allowed to define what constituted 'urban' and that this is unlikely to equate with any British definition. The response of East African interviewees revealed a majority of them also to be from rural backgrounds with 58 per cent defining themselves as rural and 42 per cent urban. Such a breakdown is somewhat unusual within the overall context of East African migration to Britain, since, as was noted in chapter 3, the Asian population was overwhelmingly urban. This suggests that those East Africans who have settled in Blackburn were the poorer, less well-educated members of that community, and as such, can be expected to be less different from their South Asian counterparts than would ordinarily be the case. Nevertheless, the two groups can draw on contrasting experience in at least one major area, that of employment. The 1978 sample were asked what occupation they had pursued immediately prior to their arrival in the UK. The results have been tabulated and are presented in table 7.4. It is immediately apparent that, despite their rural background, the East Africans in Blackburn do confirm at least one of the generalizations that are frequently made about members of that group; they were overwhelmingly employed in business before their expulsion. They also underscore a second generalization which was made in chapter 3 concerning the recent emphasis which the group had placed on mobility through education; 28 per cent of the 1978 sample had been involved in full-time and further education prior to leaving East Africa but had had this interrupted by expulsion. Beyond these two categories, other important occupations included labouring (6.2 per cent), clerks (5.2 per cent), and involvement in management (4.2 per cent). In total, nearly 51 per cent could be classified as white-collar workers, 9.3 per cent as skilled manual, 3.1 per cent as semi-skilled manual, and 6.2 per cent as unskilled. The picture which emerges from the South Asian respondents is very different. Thirty-five per cent had been engaged in education prior to leaving the sub-continent, but the majority of these had been at school. They had left school and migrated to Britain as soon as they could legally look for employment here, and regarded their education as complete. Almost as large a percentage had been farmers, and to a lesser extent labourers. Other important occupations included the Armed Forces (3.4 per cent), business (7.6 per cent), and the professions (3.4 per cent). Again, these data can be aggregated to indicate that 14.1 per cent of South Asians had previously been white-collar workers, 7.2 per cent had been skilled manual workers, 3 per cent semi-skilled workers and 34.6 per cent unskilled. This breakdown is radically different to that of the East Africans and suggests that the latter might be better equipped in terms of both experience and aspirations to succeed in British society.

The same can also be said of the East Africans' knowledge of the English language. In the 1977 Asian Census, 431 out of 577 East African adult males were able to speak and write English; this represents 74 per cent, and contrasts with the South Asians where 62 per cent had the necessary skills. The East Africans, who were largely educated through the medium of the English language prior to their arrival in this country, were thus more skilled in English despite the fact that they had been in Britain for an average of 7.4 years less than their South Asian counterparts. The 'average' South Asian had entered Britain during July of 1963 whereas the 'average' East African migrated in January of 1971. This differential in linguistic ability can also be shown to follow through to those members of both groups who are at school in Britain. Robinson (1980c) was able to collect a sample of reading ages and standardized maths scores for 627 ten-year-old children drawn from twelve schools. Common tests had been administered in each of these schools under identical conditions in order that results should be strictly comparable. Robinson (1980c: 149) concluded from his analysis of these results that 'Asian pupils were markedly over-represented in the lowest reading-age categories (7 years 6 months to 10 years 3 months), and markedly under-represented in the higher ability categories (10 years 4 months to 12 years 6 months). The mean reading age for white children (after standardization) was 10 years 6 months, whilst Asian children averaged 9 years 4 months. The same findings were replicated for standardized maths scores. The average maths scores for white children was 92.82, whilst Asian pupils average 86.04'. Robinson went on to discuss the causes of this under-achievement and mentioned, in this context, factors such as the ability of parents to speak English, residence in ethnic core areas, the social class and background of parents, and the socio-economic profile of the areas of residence. A second look at the same sample of reading ages allows an extension of this analysis to consider whether differences exist between the achievement of South Asian and East African children in Britain. It would appear that they do. The South Asians scored an average reading age of 8 years 9 months whilst the East Africans scored 9 years 3 months. Again one must recall that these differentials have appeared after less than five years of formal education. They suggest that the advantage which East African adults enjoy in terms of their ability to speak and write English is being passed on to their children who should consequently find themselves in a more competitive position when they enter the labour market.

However, adult East Africans do not simply possess greater knowledge of English, for they have also been in receipt of lengthier formal education which is reflected in the qualifications which they have gained. Analysis of the responses of those 2450 adult males interviewed in the 1977 Asian Census allows a comparison between the proportion of East Africans and South Asians who possess formal academic or vocational qualifications. Four hundred and forty East Africans possessed no qualifications whatsoever, whilst 137 declared that they had achieved at least CSE level. Twenty-four per cent were therefore qualified in some way or other. 1568 South Asians were unqualified, whilst 305 had qualifications. In other words, only 16 per cent had been educated to CSE level (or its Indian or Pakistani equivalent) or beyond.

Summary

Despite the fact that those East Africans who have settled in Blackburn appear to have been less privileged prior to migration than most other East Africans in Britain, the Blackburn data have demonstrated that clear differences exist between them and the South Asians in the town. The motives behind the migration of South Asians, and its timing and destination, all indicate that Indians and Pakistanis came to Britain as economic target migrants who sought to raise the social standing of their families back home. Only those who were able to make an economic contribution came to Britain during the early phases of migration, and this still has implications for the age and sex structure of the minority in Blackburn, despite the process of family reunion. The purpose of migration has also influenced the attitude of migrants towards participation in British society. They have a 'sojourner' mentality (Siu, 1952) built around a belief in return migration. Moreover, the tendencies which these factors produce towards encapsulatory, rather than usurpationary, closure are strengthened by the limited potential of the migrants' backgrounds. Even if South Asians were keen to participate in usurpationary closure against the white working class, in Blackburn they are singularly ill-equipped to do so.

The conclusions regarding East Africans are somewhat different. They were largely expelled from their previous countries of residence and many came to Britain because they had nowhere else to go. Because of the circumstances of their expulsion they came as communities to Britain where they form a demographically more stable and complete population. This stability should be enhanced by the gradual arrival of other dependants who were accepted by India on a temporary basis during the varying crises. The quota system is, however, preventing more rapid family reunion. The East Africans also suffered the disability that their migration was unplanned and, in many cases, accomplished without significant resources. However, the East Africans came from a background which is likely to stimulate and allow effective usurpationary closure in an effort to gain access to societal resources.

CHAPTER 8

The Empirical Operation of Structural Forces and Cultural Preferences

This chapter is concerned with the relationship which exists between the white indigenous population of Blackburn and the town's Asian minority. The analysis therefore assumes that 'the' Asian population in Blackburn is homogeneous and coherent and that all members are engaged in similar relations with the white population. Subsequent chapters will, in practice, demonstrate that this assumption is fallacious and that it is more accurate to think in terms of the Asian populations of Blackburn. However, since this particular chapter concentrates upon the macro-structure within which *all* Asians are forced to operate, and since whites rarely differentiate between a Punjabi Sikh or a Gujurati Hindu, it would seem that such an assumption is reasonable for this phase of the analysis.

The material presented in this chapter illustrates the extent to which Asians are seen both by the indigenous working class and middle class as a group who can legitimately be regarded as being beyond the normal bounds of society. It also assesses the extent to which Asians are denied access to scarce resources by whites through a system of stereotyping and exclusionary closure. However, exclusionary closure only establishes potential boundaries and limits, whereas cultural preferences determine the degree to which these limits are challenged. The second part of the chapter therefore investigates the attitudes of the Asian population towards British urban society and whether their defensive reactions and cultural preferences stimulate them to participate in usurpationary closure. Lastly, the strength of structural forces and cultural preferences are assessed in order to determine whether the two groups have a competitive relationship based on conflict, or a *modus vivendi* of avoidance and complementarity.

Structural Forces

Attitudes, stereotypes, and reactions

The task of studying the indigenous population of Blackburn is made considerably easier by the existence of Jeremy Seabrook's *City Close-up* (1971), a most detailed and revealing text which presents a history and sociology of the town as seen through the eyes of its inhabitants. Seabrook's material bears heavily upon the issues which are the concern of this chapter, notably the attitudes of whites to 'the immigrants' in the town. It is perhaps best to let this material speak for itself, as Seabrook did, rather than re-organize it into preconceived categories and frameworks. In any event a clear pattern of consensual views arises naturally:

Evelyn Watson, 63 year old social worker: 'We get the white girls who run after the immigrants—for some reason unknown to me in my old age. They've got some strange fascination for them, because they do run after them.' (p. 25.)

John Johnson, 63 year old reedmaker: 'For the home trade they were very strict, but for what they called ''for th'niggers'' wasn't so strict. For the stuff that went abroad it was a question of production, production, as much as you could, but for the home trade it was class. . .If they live next door to me I'll bring them up to my level, I won't let them drag me down.'

John's neighbour: 'There's this chap. white chap, laying in bed and he hears a noise, up in the roof, and he thinks to himself, ''Happen there's rats or something up there.'' So he goes through the trapdoor like, that leads into the roof and he sees a great long row of mattresses in the roof, all the length of the street, and on every one is a Pakistani. Come in through the roof from a Pakistani house down the other end of the street.'

John: 'Before I believe that I'd have to see it myself. . .Because what I've seen up there is lumps of plaster and slates. . .They'd only have to put a foot wrong to come through the ceiling. . .But for anybody sleeping up there, I don't think they could do it. . .There isn't room to put a mattress up there without coming through. . .I heard a tale the other day about a Pakistani who took his car to be repaired, and when he went for it it were £40, and he said he hadn't £40. So he went round to Social Security, and he told them about it, and he took two or three of his mates to work in it. And Social Security gave him £40 for his car being repaired.'

Neighbour: 'I know somebody and he worked on nights in this mill in Blackburn, and three or four nights in succession there were 54 meals missing of a night time, 54 of what they'd planned for the night. And the night after it happened again. So the next night they had a round-up, the security police. And they found 54 Pakistanis asleep on top of the belts—their mates had brought them in to give them a night's sleep and the meals.' (pp. 34–5.)

Labour party supporter: 'A lot of people are scared of saying this. Well, I'm not scared of saying it. A lot of my friends are all socialists and we all work for the Labour Party, but they're scared stiff of saying that Enoch Powell to a degree is right. . .Look what happened in 1947 in India. I bet there's one or two chaps here now that fought out in India. They kicked us out, send the British home, we don't want them. . .I'm having to work now to give my children a decent standard of living, and these immigrants are a threat to what I'm trying to do.' (p. 41.)

Mrs Frost and neighbour: 'Another thing is that most of us are getting alarmed at the amount of numbers that are being allowed to congregate and it's starting to get a ghetto . . . In the street that I live in, there's 42 houses, of which only 17 are occupied by white people now. The rest have been, the word is driven out. If we do come to sell our property we're offered a quarter of what it's worth by view of the fact that it's coloured people, that there's no one else to come into them. I've been in some of these houses and the interior of their houses are clean, but they are certainly so far behind us in our way of life that it's unbearable.'

'Well, Blackburn Street: a few years ago it was in the Evening Telegraph that it was one of the nicest streets in Blackburn. . .Well I've got a grating in front of my house, because I've got a cellar, and I've seen a little boy weeing down it, into our houses what we're living in. It isn't very nice, I've lived there for 28 years and it's been, well, a lovely residential area. . .it's nothing but a slum now. I speak the truth, it's filth. . .And we're afraid of disease, we'll soon get a epidemic or something I'm sure. . .In this day and age, when science is getting all sorts. . .fascinating, we're going back to the Stone Age, where they can do as they like on the pavement. And if they want to live like that they can go back where they came from.'

'How many white people living in this district wasn't followed during the winter, in the darkest months, by dark people, because I were.'

'Why are they allowed to get the Social Security and the child allowance. . .when they've never paid anything into our country? Look at me, I've got 26 years' stamps on, but I can't get a pension at 60. . .Now then, I don't get a pension, but they get it handed to them on a plate.'

'They're outnumbering us because a lot come illegal. They come in the night.'

'In a few years, the way they're breeding, they'll take over.'

'Once upon a time if you were in any trouble or you needed help, you could go to anybody in the street. Can you now? I don't think as Lily and I ever speak to anybody hardly.' (pp. 44–57.)

Roy Sharratt, 27 year old bingo caller: 'The country seems to be doing more for them than what they're doing for its rightful people who were born here. We should be first and they should come second. They should come last in fact . . . You sell a Pakistani a house and then watch the price of the house next door get devalued.' (p. 63.)

Fred, 47 year old fruitstall-holder: 'It's all right saying we've exploited this country, we've exploited that country, but you can't tell me one country which since we gave them home rule, has done any good. . .Oh, they'll come to us for another twenty thousand, another fifty thousand. . .They're worse off now, Pakistan and India, than they've ever been. So never mind we exploited them.' (pp. 69–71.)

Mrs Grayson, businessman's wife: 'But the type of immigrants we seem to be getting in Blackburn, they're getting good jobs in the cotton mills. . .in buses, and things like that, and they do not seem to me to be contributing one single thing to the welfare of the town. . .I have a friend whose house had dropped in value from £1500—a small, respectable, beautifully kept terraced house. The immigrant families are infiltrating, and she cannot now even get £500.' (p. 80.)

Edward Poynton, 70 years old: 'I mean, it's no use, you can't pretend to set a primitive savage in the middle of a civilised community and expect him to react in a way that another civilised person would have done. He's not used to the amenities, he doesn't understand them. He's out of his depth right away. . .They need some form of training. . .It could be a School of Civilisation.' (p. 90.)

Rosemary Bragg, 29 year old mother of six: 'I'm not prejudiced against the immigrants, but they all seem to dress their children nice. They seem to get these houses, and they've jobs to go to. They seem to make more money than us altogether. . .they seem to be better off altogether than the English people.' (p. 107.)

Mr Morley, retired: 'Well, as I told you, I remember the natives from other countries coming to help us out during the war.' (p. 263.)

These brief excerpts from the more extended views of Blackburn's white residents are by no means atypical of others which Seabrook quotes. Even these extracts though, underline several important points about the way in which Asians are viewed by Blackburnians, young and old, working and middle class. Perhaps the most obvious point which arises is that the immigrants are without doubt viewed as a threat. Robin Ward's suggestion that inter-racial attitudes in Sparkbrook may have resulted from the threat which blacks posed to *neighbour-hood* status seems to be only a part truth. Rather, a much larger complex of attitudes appears to exist which result from the threat which immigrants are thought to pose to *all* facets of life. The quotes from Seabrook illustrate that 'the immigrants' are cast in the role of a threat to almost everything: the moral fibre of local women; decent standards of housing; the respectable life-style of the working-class Blackburnian; property values; full employment; the British sense of natural justice as embodied in the welfare system; the standard of living; the health of the white population; the desirability of their residential areas; law and order; the traditional working-class pattern of social interaction; and ultimately control of the State. Secondly, and not entirely independent of the first point, 'the immigrants' are also seen very much as active competitors for a wide range of scarce resources, be it housing, jobs, women, unemployment benefit, or a good standard of living. Finally, whites in Blackburn do seem to regard Asians as inherently inferior to themselves, an attitude which, on closer analysis, has its origin in the language and myth of the colonial era. Words such as 'savages' and 'natives' became attached to Asians, as did notions of 'dragging them up to our level', and making them 'civilized.' John Johnson appears to be a case in point. Although he holds some superficially liberal views, when he is told the story of Pakistanis who had occupied the lofts of a whole row of terraced houses, he says he cannot belive this. But, it transpires that his motive for disbelieving the tale is not that Pakistanis would be unlikely to do such a thing, but that the ceiling of a terraced house could not support the weight of so many people. He also demonstrates how, during colonial rule, even the textile trade embodied the sentiment that second best was good enough 'for th'niggers'. Inevitably, such sentiments would be internalized within the value-system of those who worked in the trade. Further links with the colonial era and its associated attitudes are revealed in the Labour Party supporter's comments about being 'kicked out' of India and fighting the Indians during the Independence struggle. Fred, the fruitstall-holder, also harks back to colonial days and even attempts to legitimize colonial rule by explaining that it was for the good of the local population who were incapable of looking after themselves.

Seabrook's evidence suggests that many of the whites who live in Blackburn do have those attitudes towards Asians which the model presented in chapter 6 would predict. They do see Asians as competitors for scarce resources, and a threat to many facets of traditional working and middle-class life. And they do relegate them to a position which is firmly outside mainstream society, and hence legitimizes their treatment as inferiors. Finally there is some evidence to suggest that these attitudes have their roots, not only in competition, but in the colonial system. However it is important to ascertain whether this complex of attitudes produces overt action or whether it remains latent.

One clear embodiment of such ideas is the support afforded to extreme right-wing, quasi-political parties such as the National Front or National Party (see Walker, 1977, Taylor, 1982). Whilst one could perhaps regard a vote for one of these parties as merely a protest vote against the policies of the major parties (Husbands, 1975), it is nevertheless a positive act which often requires premeditation. It cannot therefore be dismissed lightly. This becomes even more true when one considers that a vote for the National Party in Blackburn would mean a vote for either John Kingsley Read (who became infamous for his 'one down, one million to go' speech on the death of an Asian youth in Southall during June 1976) or people of the ilk of Mr Horman, the chairman of the local branch, who was happy to write a letter to the local paper which contained the sentences, 'We, as everyone in this town well knows are a racialist political party' and, 'The National Party should be invited to give the meeting the balanced effect it will require to create peace until the time comes for the human repatriation of the coloured immigrants' sic (Lancashire Evening Telegraph, 1976). Despite the fact that the party's platform was openly racialist (or perhaps because of it), the National Party polled 1588 votes in St Jude's Ward during May 1976 (Lancashire Evening Telegraph, 8 July 1976). This proved to be 19 per cent of the total vote, a figure sufficient to ensure the election of two National Party councillors, the first elected representatives of their party in Britain. Moreover, Walker (1977) points to the fact that Blackburn was one of the few National Front branches able to raise its own election deposit without recourse to central funds. When one takes this in alliance with the success that the National Front had enjoyed in the May 1972 local government elections, the April 1973 county elections, and the June 1973 district elections (Le Lohé, 1976), one is left in little doubt about the attitudes of many people in Blackburn towards the presence of Asians in the town.

Spatial manifestations of the power structure

The individuals within Blackburn who control access to housing, jobs, and the other elements of societal resources are not barred from their position if they hold firm negative views on Asian immigrants, in the same way that they would not be barred if they had a deep conviction about nuclear disarmament. They are able, if they so wish, to operationalize their convictions either openly or in a manner likely to conceal the true purpose of their actions. Clearly the impact of the Race Relations Act is likely to have shifted the emphasis towards the latter, although it is unlikely to have eradicated the former. Given that such attitudes are widely held in the local population and may even permeate the coterie of local decision makers, it is important to determine whether they produce behavioural correlates and therefore spatial and social manifestations. Again, housing provides the case-study by which these notions can be tested.

Even a cursory analysis of residential segregation in Blackburn reveals that there is spatial separation between the white and Asian populations, and that this is present regardless of the scale or method of measurement. The 1971 Census provided a useful starting point from which to quantify residential segregation, in view of the fact that previous censuses had contained substantial errors through underenumeration and misenumeration (Peach and Winchester, 1974). Use of census data allows the calculation of Indices of Dissimilarity (see Peach, 1975 for a general description and Duncan and Duncan, 1955 for a detailed technical exposition) at ward and enumeration district (ED) level, whilst the electoral roll provides a source of information at street level, always assuming of course that one regards the identification of ethnic Asians by their names to be a reliable procedure. In 1971, then, the Index of Dissimilarity (ID) between Asian and white individuals at ward level in Blackburn was 53.6 whilst similar calculations at ED level produced a value of 73.6 (Robinson, 1981a). At the street level, using households instead of individuals, the ID rose to 85.3. Bearing in mind that the theoretical range of the ID is zero (an absence of segregation) to one hundred (complete segregation) these values represent significant segregation. Expressed another way, almost nine out of every ten Asian households would have had to change their street of residence to produce a distribution pattern similar to that of the white population. Clearly, segregation at this level, which Jones and McEvoy (1978) maintain is the most realistic one for measurement in the British context, was massive. One alternative method of demonstrating the spatial separation of the two groups is the location quotient (see Lee, 1977 for a general description of this measure). The location quotient is a means of quantifying the over- or under-representation of minority groups in particular areas compared with the distribution of the total population. A value of between zero and one is indicative of under-representation, and values greater than one signify over-representation. However the location quotient can also be used as a measure of city-wide concentration rather than simply the over- or under-representation of a group in one of the city's constituent sub-areas. This use of the location quotient (LQ) was first pioneered by Doherty (1969). In 1971, in Blackburn, the range of ED LQs was from 0.03 to 10.9, and the average ED had a calculated LQ of 1.7. Stated in a different way, the average Asian lived in an ED with an LQ of 5.9. Asians resided in only 117 of the town's 210 EDs (55.7 per cent of the total). Table 8.1 demonstrates the exact breakdown of the LQ data and indicates how over half the Asian population lived in EDs where they were markedly over-represented whilst only 7.3 per cent lived in areas of under-representation. Finally, a third measure of segregation can be employed to add further detail to the description of group separation. This is P*, first proposed by Bell (1954) and subsequently resurrected by Lieberson (1981). P* has the advantage that it is capable of measuring group isolation in an asymmetric manner since it takes into account the relative size of the two groups involved in the calculation. For the purpose of the present analysis it is probably of more value to consider the isolation of an average Asian from other members of the ethnic population ($_aP^*_a$), since this index functions in a complementary manner to the ID and describes a different facet of group separation. Calculation of a $_aP^*_a$ value, at ED level for Blackburn in 1971, thus reveals that the average Asian lived in an ED where approximately 28 per cent of the population was also Asian. This compares with the 9 per cent value which would have resulted from an evenly-distributed Asian population.

Regardless of whether one used the ID, LQ, or P^*, it appears from the 1971 Census that the white and Asian populations were indeed markedly separated in residential space at the beginning of the decade. However, before going on to discuss the exact form and nature of this separation one must be sure that it was not a temporary phenomenon induced by conditions unique to that era. Diachronic analysis is therefore called for. Table 8.2 thus provides ward level IDs calculated from the electoral register for the years 1968 to 1984 whilst table 8.3 gives time series data for $_aP^*_a$ at the same level from 1968 to 1984. The former indicate that residential dissimilarity increased from 1968 to 1973 and has fluctuated since then around the mid-fifties. There is some evidence of a decrease since mid-decade but this would be difficult to prove conclusively because of changes in ward boundaries. However, when one takes into account the numerical increase in the Asian population of the town (as P^* does), the average Asian has in fact become progressively more residentially isolated throughout the period, at least at ward level.

The 1977 Asian Census allows a more detailed analysis of changes in group separation in the six-year period 1971–77. At ward level, the ID between the white and Asian population had fallen slightly to 51.0, although at ED level the decrease could not be proved significant (73.6 in 1971 to 73.3 in 1977). At the finest scale of analysis, the street, the ID had fallen by 10 points to 75.3. In terms of residential dissimilarity, then, the six-year interval was characterized by a decrease, although it is important to note that the levels recorded for 1977 were still substantial in their own right. Moreover, similar conclusions arise from the LQ analysis, the results of which are presented in table 8.1. These point to continuing high levels of separation (nearly 52 per cent of the Asian population still found in EDs with LQs of greater than 5) although it is note-worthy that there has been an increase in the percentage of the Asian population living in areas of lesser over-representation (LQs of 1 to 3). Again, though, the P^* values suggest that the decline in residential dissimilarity between Asians and whites has been more than compensated for by the numerical increase of the former. In 1977, the average Asian lived in an ED where over half the other residents were Asian, a significant increase on the 28 per cent value recorded for 1971 (see Robinson, 1980e for a discussion of this point).

All the measures employed here point to the fact that the Asian population cannot be described as residentially integrated. In 1977, Asians were still only found in a little over half the EDs within the town, and 58 per cent of the minority resided in only eleven of the 210 possible EDs. In the light of the housing-class model and the evidence presented at the beginning of this chapter, there seem to be good *a priori* reasons for assuming that the separation of the white and Asian populations results from actions of exclusionary closure by the former. However it would be difficult to support such an argument if the

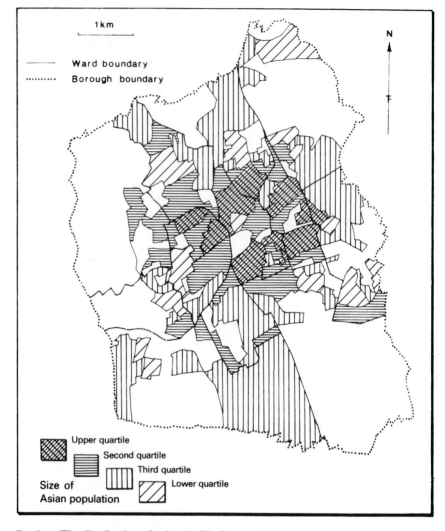

FIG 8.1:　The distribution of Asians in Blackburn, 1971

Asian population could be shown to enjoy good housing as a result of gaining access to societal resources.

Whilst aggregate measures of group separation provide a good description of the extent of residential mixing, they do ignore the actual location of areas of concentration or under-representation. Figures 8.1 and 8.2 make up for this criticism by providing a graphical representation of the distribution of the Asian population in 1971 and 1977. The distribution is clearly dominated by an inner-ring of core areas which encircle the central area of the town in all sectors except the south-west, which is characterized by non-residential land-use and council estates. In certain places the inner-ring also has extensions which reach out to the suburbs, although the latter are themselves devoid of substantial Asian settlement. More importantly, perhaps, one can describe the general socio-residential characteristics of this inner-ring by reference to the cluster analysis presented in Chapter 2. Of those areas of over-representation (LQ of 1.0 or greater) 13 were classified as belonging to cluster 52, 12 to cluster 22, 6 each to clusters 21 and 23, 3 each to clusters 25 and 26, 2 each to clusters 27 and 33, and 1 to both clusters 19 and 51. A complete socio-residential profile of each of these clusters is provided in Appendix I, but suffice it to say that they are generally areas of poor or very poor terraced housing which lack amenities. In short, areas of deprivation.

An issue of some importance here is the ecological fallacy (Robinson, 1950) which plagues most analyses based on aggregate census data. Whilst the census

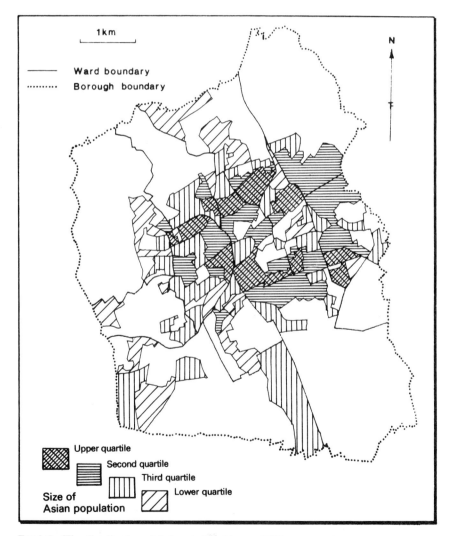

FIG 8.2: The distribution of Asians in Blackburn, 1977

allows researchers to identify in which areas a target population lives, one cannot necessarily impute the general characteristics of that area to specific portions of the population which live within it. However the 1977 Asian Census and the 1978 Sample Survey allow a consideration of the characteristics of the Asian population in isolation from others who might live in the same area. It is therefore possible to ascertain whether Asians form a deprived group or whether they are a group who live in deprived areas.

The 1978 Survey enquired initially about the type of accommodation, and discovered that 98.4 per cent of heads of household resided in terraced properties, leaving only 0.8 per cent in semi-detached and 0.8 per cent in detached properties. These figures contrast markedly even with those for the total coloured population of Britain as revealed by the contemporary General Household Survey (1980). The latter demonstrated that, at a national level, 4 per cent of coloured people live in detached houses, 17 per cent in semis, 46 per cent in terraces, and 31 per cent in flats of differing types (GHS table 3.37, p. 48). The discrepancy between the coloured population at a national level and the Asian population at a local level in Blackburn demonstrates not only that the latter is relatively more disadvantaged but also that the housing stock in Blackburn is still made up largely of properties erected during the Industrial Revolution. This also has repercussions for other housing characteristics such as housing size. The 1978 Survey produced figures which show that the accommodation of the average Asian household in Blackburn consists of 5.76 rooms, of which 2.55 are bedrooms. Since the mean Asian household size in Blackburn is 5.23 persons (1977 Asian Census), this gives an average of 0.91 persons per room, a figure which exceeds that for the total population of the town. The age of properties is also reflected in the access which Asian families have to housing amenities: according to the 1977 Census, only 74 per cent of households had access to an installed and plumbed bath. This compares with the GHS's finding that, in Britain as a whole, 96 per cent of coloured households had sole or shared use of a bath (GHS table 3.41, p. 49). Finally, the location, size, and quality of Asian-occupied properties have their corollary in a low market value. The average from the 1977 Sample Survey was only £4440.

When David Smith (1974) published the second PEP report on racial disadvantage in Britain, he commented upon the paradox that, whilst amongst the indigenous population owner occupiers enjoyed better housing than other tenure groups, the converse was true amongst the Asian population. He argues that, for Asians, owner occupation was a means of getting cheaper housing, not better housing. This seems also to be true of Blackburn. Asians there live in small, cheap, and obsolete houses, but the 1977 Asian Census also demonstrated that they are overwhelmingly owner occupiers. Over 94 per cent of heads of household proved to be in this tenure group, whilst 3.1 per cent were renters of furnished property, 1.2 per cent of unfurnished property, and 0.5 per cent of local authority accommodation. Table 8.4 contrasts this tenure breakdown with others derived from national surveys at around the same time. It is clear that if owner occupation is more frequently linked with poorer accommodation, then the Asian population of Blackburn appears to be unusually disadvantaged yet again. Clearly, the limited opportunities which the housing stock offers for renting have an important role to play here.

Knowledge of the attitudes of whites in Blackburn to those Asians who live in the town would again suggest that the latter remain locked into poor housing in declining inner areas because of the exclusionary closure of the former. Sceptics could argue, however, that Asians choose to spend their income upon consumer durables and a better life-style than their white counterparts and that this affects their ability to provide better housing. The 1978 Survey indicates that such a line of argument is not justified. Questions were deliberately asked in the survey to allow comparison with the results of the 1978 General Household Survey (1980). Thus, whilst 36 per cent of Asians in Blackburn possessed a colour television, 44 per cent of all coloured households in Britain did so. Other comparisons were as follows: ownership of car, 37 per cent in Blackburn, 45 per cent nationally; ownership of refrigerator, 80 per cent in Blackburn, 85 per cent nationally; and possession of a washing machine, 39 per cent in Blackburn, 45 per cent nationally. The results of these direct questions also received support from an open-ended question about the relative advantage of banking with Asian or British banks. The most common reply to this was that respondents had no experience on which to base an opinion since they had insufficient money to have ever had a bank account.

Poor housing conditions and the absence of evidence of a materially-oriented life-style suggest that explanations for residential disadvantage must come from other quarters. The most obvious of these is the presence of discrimination in the labour market and, at a later stage, in the housing market. Clear evidence to show that this exists is not lacking at national level (see chapter 4) and both the Asian Census and Sample Survey provide data which suggest that a parallel situation exists at a local level. The labour market was considered first. According to the Asian Census, the rate of unemployment amongst Blackburn's economically-active Asians was approaching 21 per cent in 1977, at a time when the comparable figure for the town's whole population (including Asians) was barely 3 per cent. Even disregarding this higher level of unemployment, which could be a product of linguistic difficulties, there seemed to be a strong case for arguing that the Asian population had only been allowed to act as replacement labour, thereby gaining access only to those sectors of the local economy which had been progressively abandoned by the indigenes. This is reflected in the industrial structure of Asians in Blackburn (see table 8.5), which shows a marked over-representation in those areas now considered undesirable by whites either because of irregular hours, low pay, or the physically demanding nature of the work. Sixty-four per cent of respondents were consequently employed in either textiles, paper making, transport, metal manufacture, or the chemical industry. In each of these cases, local employers have difficulty filling vacancies, and in some the struggle for labour has even generated fears about the local existence of the industry itself. However, the converse of Asian concentration in sectors such as textiles is that the group is markedly under-represented in the higher-status occupations which society recognizes as desirable: only 64 respondents in the Asian Census recorded that they held posts in the professions, finance, or public administration. Moreover, the notion that Asians act as replacement labour receives further support when it becomes clear that, even within industrial sectors, they are found in the jobs with the lowest status and poorest prospects. Only 4.1 per cent of respondents in the 1977 Census could be classified as members of either social class 1 or 2 (professional and intermediate), whilst 33.9 per cent belonged to social class 3 (skilled), 48.9 per cent to social class 4 (partly skilled), and 13 per cent to social class 5 (unskilled). The corollary of this is that income levels are generally low despite the prevalence of overtime and shift working. The average weekly income for Asian heads of household in Blackburn who were in full-time employment was thus £51.68, or an annual income of less that £2700. This was supplemented by an average of £9.32 per week from other sources such as family income supplement, social security, or a working wife's income. In total then, the average household income was £3172 per annum at 1978 values i.e. approximately 62 per cent of the national industrial average (*Department of Employment Gazette*, 1979) and 68 per cent of the national average for workers in the textiles industry (same source). What is more, these earnings were the product of a working week which averaged 41.1 hours, and was more often than not made up of shiftwork (less than 25 per cent of the 1978 sample did *not* work in some form of shifts). It hardly seems surprising in the light of these facts that when respondents in the Sample Survey were asked what they liked or disliked about their jobs, several simply stated that it was the only job that they could get. One man went further than this and commented on the fact that employers expected Asians to work much harder for the same wage than an equivalent white employee. Comments such as these raise the issue of direct or institutional discrimination. Of those 264 respondents in the 1978 sample who had a firm opinion on the matter, 62.5 per cent felt that they *were* treated differently in Britain solely because of their colour, whilst nearly 20 per cent of the total sample said they experienced racial discrimination in their everyday lives either 'frequently' or 'very frequently'. Put another way, when asked to specify those things which made them feel insecure about life in Britain, 57 per cent named the National Front, 15 per cent 'the racial problem', 4.1 per cent personal security, 5 per cent the policies of the major political parties towards ethnic minorities, 2.3 per cent the fear of involuntary repatriation, and 1.8 per cent racial discrimination. It is clear that the issue of racial discrimination, in its broadest sense, is uppermost in the minds of many Asians in Blackburn, and that, rightly or wrongly, at least some hold a perception of British society in which the local population is characterized as being overtly hostile. The attitudes which Seabrook (1971) uncovered in Blackburn lend considerable credence to this view and the 1978 Survey demonstrates how such perceptions are translated into behaviour. Over 26 per cent of Asian heads of household admitted that they had no white friends whatsoever, whilst a further 9.2 per cent said they had 'hardly any' and 41.1 per cent had 'a few' (cf. Kannan, 1978: 155).

The same issues of racial disadvantage and racial discrimination also follow through into the residential sphere and the question of housing provision.

Insecure employment and low incomes debar many Asians from building society mortgages and bank loans, and they are therefore forced to turn to what Rex and Moore (1967) termed 'illegitimate' methods of housing finance. In Blackburn, during 1978, only 6.6 per cent of the sample were using building society loans to finance owner occupation, a figure which no doubt also reflects the societies' unwillingness to become involved with older properties or ones with short leases. Potential Asian owner occupiers are therefore forced to raise loans either from finance companies (3.3 per cent), friends (1.5 per cent), or relatives (9.6 per cent), or, as more frequently happens, they buy cheaper houses outright with the resources of the joint family (32.2 per cent). Other methods of owner occupation include private mortgages (6.3 per cent) arranged by community brokers and the socially-orientated local authority mortgage scheme designed specifically to aid low income groups who wish to buy inner-city properties (26.2 per cent). In short, because of their position in the labour market, few Asians can become owner occupiers through the 'conventional' channels used by their white counterparts.

Moreover, the effects of racial disadvantage are also compounded by a variety of forms of racial discrimination in housing markets. Both estate agents and vendors are known to discriminate, and it seems unlikely that this is not the case in Blackburn. Racial discrimination significantly alters the search behaviour of Asians in Blackburn who are looking for new property, for they rely to a much greater extent upon friends and relations (34.2 per cent), and informal channels such as private advertisements in houses (19.3 per cent), shops (21 per cent), and local newspapers (1.8 per cent). Furthermore, where they do use the services of an estate agent, the latter may himself be Asian (Robinson, 1979b), and therefore unlikely to discriminate against members of his own group. A small number of families have also bought property on new estates, another acknowledged way of avoiding many discriminatory practices. However, even for those who have successfully found a property that suits their needs, there is occasionally the problem of a vendor who is unwilling to sell to Asians either because he fears the displeasure of neighbours and friends who must remain in the street, or because he does not want to see 'his' house owned by coloured people. In certain cases vendors have their prejudices overcome by Asians who are willing to offer more than the asking price, but in others they remain resolute. A case in point is 23-year-old Graham James who lived in Norwich Street in Blackburn. When he decided to sell his house he erected a 'For whites only' sign beneath the 'For Sale' notice, and despite coverage in the local press refused to back down for some considerable time. Ultimately he was persuaded to remove the sign by friends who feared that he would soon be prosecuted under the race relations legislation. When interviewed by the local *Evening Telegraph*, Mr James said, 'They were trying to force me out, and have offered £1500 less that the sum I have mentioned. I did it because I would not be forced out'. Clearly Mr James felt that he was the passive party simply reacting against unfair pressure. In those cases where Asians *have* persuaded whites to sell to them, there is then the added complication of the attitude of neighbours to the arrival of coloured families in 'their' street or area. As has been shown by the American literature (Rapkin and Grigsby, 1960), the reaction of whites to this event is determined to a large extent by their proximity to what they consider to be 'the ghetto'. The range of responses is again exemplified by occurrences within Blackburn. Robinson (1981b) has demonstrated that in inner areas of terraced property the attitude of whites to the arrival of Asian families is characterized by four major stages (see fig. 8.3). Robinson labelled these 'reaction', readjustment', 'transition', and 'withdrawal'. He described 'reaction' as follows:

(i) 'Reaction' is dominated by the fears of white residents after the arrival of the first Asian family in the street. The fears are formulated in terms of stereotypes and rational-ised by reference to falling property values and declining neighbourhood status. Poten-tial white in-movers withdraw their demand in panic. . .(and this). . .produces a fall in property values (when standardised for quality and the effects of inflation). (Robinson, 1981b: 20–1.)

'Reaction' was thought to last two years and because of low demand from whites for houses, Asians were able to consolidate their original presence. 'Reaction' was followed by 'readjustment', which

is characterised by the subsidence of panic and a return to more rational thinking. The percentage which Asians form of house purchasers falls from 40 to 10, and Asians are forced by renewed white demand to purchase property at a higher price. The percentage which Asians form of the street population rises to between 8 and 10. (Robinson, 1981b: 21.)

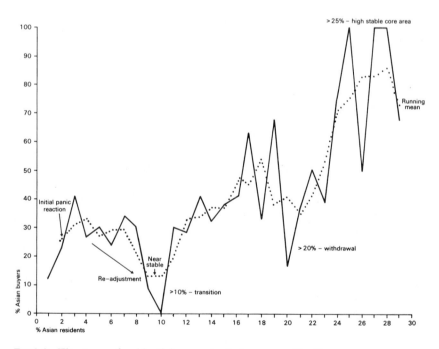

FIG 8.3: The stages of residential succession in inner-area Blackburn

However 'readjustment' was short-lived and soon gave way to 'transition'.

'Transition' is triggered when Asians form more than 10 per cent of a street's population. The area becomes less attractive to white buyers who perceive a decline in status. Consequently, the proportion that Asians form of house purchasers rises gradually to over half, and prices fall. (Robinson, 1981b: 21.)

Finally, 'withdrawal' takes place and the street enjoys 'stability' as an Asian core area. 'Withdrawal' was triggered when Asians formed greater than 20 per cent of the street population and

potential white buyers withdrew *en bloc*. The percentage that Asians form of house purchasers consequently rises to 100 within a period of approximately one year, whilst the percentage which Asians form of the street population also rises sharply. Asian demand stabilises or forces up property values. (Robinson, 1981: 21–2.)

Robinson (1981b) concluded on the basis of his data that the process of transition appeared relatively rapid and largely irreversible, and that the growth of the Asian population in a street beyond 10 and 20 per cent of the total produced non-linear changes in the behaviour of the white population. He argued finally that

the factors stated above, when taken in conjunction with the characteristics of the various phases of transition, suggest that whites are not tolerant of racial mixing in residential neighbourhoods;. . .the initial panic reaction, and the later phased withdrawal of white residents and buyers point to a desire on the part of whites to minimise residential proximity to Asians. (Robinson, 1981b: 23.)

Robinson's conclusions suggest that, in the areas of terraced housing near the town centre (which are therefore near to existing Asian core areas), whites eventually abandon streets to Asians after a very temporary period of resistance.

The same cannot be said for those areas more distant from the town centre, and therefore more distant from Asian core areas. One of the major concerns of the community relations staff during part of 1978 was the issue of a family who had moved into a previously white street in an area of above average housing. During the first month of its residence there, the family was forced to endure verbal abuse and had its windows smashed several times. It was considering moving out as a result. In this case then, the white residents must have felt that they needed to resist Asian encroachment, presumably to encourage the existing Asian family to leave, and prevent others joining them. A flavour of this kind of behaviour can be gleaned from the comments of another Asian respondent in the 1978 Sample Survey who told the interviewer that, 'Hooligans come into the street and knock on our door at midnight. They come in a crowd and call us Paki etc. I am nervous.' The fear of this kind of treatment must inevitably influence the behaviour of Asians who consider breaking out of the areas which white society allots to them.

Summary

In this section, a range of subjective and objective material has been presented to describe the way in which the attitudes of members of the host society are translated into structural forces designed to ensure exclusionary closure against Asians. On the basis of this evidence it is difficult not to conclude that the white members of British society practice substantial and significant exclusionary closure against Asians, and that this is designed to prevent access to desirable or scarce houses, jobs, and neighbourhoods. However, whether this produces inter-ethnic conflict depends also upon the attitudes and aspirations of the Asians. If the Asian population retains a different value-system which does not regard those items which whites see as desirable in the same light, then conflict may remain present but effectively latent. If the Asian population has adopted a British value-system (either working or middle class) and therefore sees the same items as desirable, then usurpationary closure by the Asians will result in inter-ethnic competition and conflict.

Spatial Manifestations of Cultural Values

Neighbourhood preference

Although the evidence suggests that, objectively, Asians are confined to the inner areas of Blackburn from whence it is difficult for them to escape, this perhaps ignores a preference on their part for neighbourhoods of this type. After all, they would not be the only group in society which prefers higher density, communal living. In fact, one of the features which comes out strongly from Seabrook's (1971) interviews with indigenous Blackburnians is their misgivings about the decline of a working-class culture which stressed communality, particularly in the residential context. Furthermore, as Dahya (1974) points out, gregarious living at a high density is very much the norm in Indian or Pakistani villages, whilst privacy and individuality are not features intrinsic to the structure of Indian or Pakistani society. It is hardly surprising, therefore, to learn that many Asians are not critical of their residence in separate inner-area enclaves (Community Relations Commission, 1977). Further, they often stress the positive aspects of such a residential milieu, as native Blackburnians did only a generation previously. When asked why they lived in the area where they did, not one respondent in the 1978 Sample Survey mentioned the lack of alternatives. The reason which was most frequently cited as the pre-eminent motive was the nearby presence of relatives (27.7 per cent), whilst other reasons proved to be the proximity of a mosque (20.1 per cent), the type and price of housing (12.1 per cent), the attractiveness of the area (10.9 per cent), and nearness to friends (10.7 per cent). The results also received support from those of the CRC (1977) in Bradford. When the focus of the analysis was changed from the *single* most important reason to a combination of the *three* most important reasons given by each individual, motives were again dominated by communality. 53.2 per cent of Blackburn respondents mentioned relatives, 46.2

per cent mentioned friends, 37.3 per cent the housing, 30.5 per cent Asian shops, 17.8 per cent nearby employment opportunities, 17 per cent the attractiveness of the area, and 11.3 per cent local health services. The attitude of Asians towards their residential environment was also reflected in replies to an open-ended (rather than structured) question which enquired about features of the neighbourhood which were liked or disliked. 58.3 per cent of the 180 people who answered the question could think of nothing that they disliked, whilst those who could tended to dwell upon minor domestic issues rather than major sources of dissatisfaction. Examples of those items mentioned as dislikes included the fact that the area was hilly, that parking was restricted because of a one-way system, that there was no telephone box, that the roads needed resurfacing, and that the houses suffered low water pressure. It is fair, however, to note that not all respondents were concerned about minor issues. One interviewee simply commented that the 'housing is deplorable, streets are filthy. The whole area is a slum', whilst another said that, 'the houses are in a terrible state'. In contrast, though, the opinion of the majority was that relatives and friends were nearby, that houses were cheap and that the neighbourhood was quiet.

A third cross-check on neighbourhood satisfaction can be derived from a separate question which probed long-term attitudes towards improvement versus relocation. As is to be expected, attitudes were not uniform but, of those 369 people who replied to the question (from the 1978 Sample), 8.1 per cent said they would like to remain in the area but that it needed no improvement, 67.7 per cent said they would prefer to remain where they were but that it should be improved, and 24.4 per cent opted to leave the area altogether. In retrospect, it seems likely that the question, which offered respondents only the options of 'moving' or 'improving', was badly pre-coded since it implied that improvement was necessary, and that the option of leaving the neighbourhood untouched did not exist. This belief is borne out by the fact that 30 respondents insisted that they be recorded as wanting to stay in their existing neighbourhood but that improvement was not necessary. Furthermore, 17 per cent of those who had opted for improvement then admitted that they could think of no worthwhile changes which could be made to the area.

However, to return to those 207 people who wished to see the environment improved and could suggest specific items, either for the houses or for the neighbourhood, the majority thought that road-related improvements were the most badly needed. More specifically, 31 per cent mentioned the roads themselves, 8 per cent the pavements, 2 per cent pelican crossings, 2 per cent street lights, and 2 per cent new garages. These compare with only 1 per cent who wished to see the houses gain central heating, 2 per cent who wished to have damp removed by remedial treatment, and 18 per cent who thought the houses should have separate kitchens rather than a rear scullery-cum-living-room.

Neighbourhood preference was also probed from a completely different angle, that of perceptual studies. The 1978 Sample was asked to consider which area of Blackburn they would most like to move to if they were given an unfettered choice. The most frequent answer to this question was that families wished to stay in the neighbourhood in which they already lived. Over 14 per cent responded in this way. The next most popular neighbourhoods were Little Harwood (13.1 per cent), Queen's Park Road (9.1 per cent), and Preston New Road (8.6 per cent). In each of these cases, the areas already contained sizeable Asian populations, and were probably not the neighbourhoods which would have been chosen by a sample of native white Blackburnians. Further, several of the popular areas amongst Asians are widely thought of as the least desirable in the whole of the town by native Blackburnians. Included amongst these are Whalley Range (chosen by 6.3 per cent of Asians) and Audley Range (chosen by 6.9 per cent), both of which are Asian core areas with objectively poor housing. It seems clear that Asians do not judge the desirability of an area on the same criteria as local whites.

The final indication of neighbourhood preference can be derived from Robinson's (1981b) analysis of property values during the different phases of residential succession. He collected data on the asking prices of properties advertised in local evening and weekly newspapers, and then used the electoral register to gain an indication of the ethnicity of both buyer and seller. He eventually had a sample which contained 180 sales of property by whites to Asians and 112 sales by Asians to Asians. Robinson introduced an element of standardization into his analysis by only considering older, poor quality, terraced housing and by standardizing prices for inflation using a component which took into account only local trends in property values. Robinson (1981b: 19) concluded from his study that Asians were sceptical of buying property in newly pioneered areas (because of the absence of community facilities and the fear of isolation and hostility), but that 'as the street becomes a stable Asian core area, a premium is demanded for residence within it. Asian buyers must compete with fellow Asians to purchase the limited number of houses that are community central. . .The action of this "community centrality premium" forces up property values. . .in the last stage of transition'. Two explanations of this phenomenon recommend themselves. Either Asians are forced to pay inflated prices simply to gain any accommodation because of exclusionary closure by whites, or, as Robinson suggests, the attractions of community living are thought sufficiently important to encourage individuals to pay a premium for residence within community territory. The former reasoning is not capable of explaining the relatively depressed prices paid by Asians to Asians for residence in newly pioneered areas, but clearly this is a topic which requires further research.

Housing satisfaction

So far, only the issue of neighbourhood preference has been considered although some strong indications have also emerged from this about housing satisfaction and preference. However, more direct evidence on this topic is available from the 1978 Sample Survey and, again, different aspects were investigated through a variety of questions. Initially, the straightforward question was asked: 'How satisfied are you with your current house?' All 364 male heads of household replied, of which 86.6 per cent were either very satisfied or satisfied, 8.2 per cent dissatisfied, and 1.9 per cent very dissatisfied. These results can be put into context through comparison with data from the General Household Survey of 1978 and the National Dwelling and Housing Survey of the same year (Department of Environment, 1978). The GHS revealed that 86 per cent of a national sample were either satisfied or very satisfied whilst the NDHS suggested a figure of 82 per cent. Eleven per cent of the GHS sample expressed some measure of dissatisfaction compared with 9 per cent in the NDHS. In the light of these figures, the levels of satisfaction found in the Asian sample appear to be very close to the national average for the white population even though Asians live in accommodation which is objectively inferior on every count. One might therefore have reasonably expected Asians to express greater degrees of dissatisfaction than they actually did. Secondly, the GHS discovered that, amongst those who were more dissatisfied with their housing, those under 30, large families, those lacking sole use of the three basic amenities, and those with insufficient bedrooms were all over-represented. Since Asians in Blackburn figure heavily in all these categories one might have again expected higher degrees of dissatisfaction than were actually apparent. This again seems to point to a different value system.

A similar conclusion arises from the question of satisfaction with the size of the property. Three per cent of the 1978 Sample thought their current house too large, nearly 80 per cent thought it 'about right', and only 17.2 per cent thought it too small. Again, by British standards, this seems a little surprising in view of the relatively high densities at which many Asian families live.

In the light of the GHS's warning about the difficulty of measuring housing satisfaction by using only relatively simple questions, the 1978 Sample Survey also contained open-ended prompts about neighbourhood satisfaction. As expected, interviewees defined 'neighbourhood' in many different ways and some talked exclusively about their accommodation rather than the broader environmental issues. These responses have already been described, and it was noted that dissatisfaction tended not be be caused by the poor structural condition of the housing or its lack of amenities but by specific, and largely individual, problems which seemed objectively to be somewhat trivial. A similar picture emerged from replies to the question about improvement or relocation. Finally, the 1978 Survey enquired about the long-term aspirations of respondents with regard both to the type of accommodation and preferred tenure. In each of these cases, a comparison of stated aspirations with present circumstances allows an indirect measurement of satisfaction. Only 1.6 per cent of heads of household thought that a flat would be ideal, whilst 95.2 per cent opted for a house. Less than one per cent aspired to private renting, 9.8 per cent to public renting, and 86.5 per cent to owner occupation. These data suggest that, in relation to their stated aspirations, Asians are slightly over-represented in the private-rented and owner-occupied sector and under-represented in the council sector. However in addition to the 0.5 per cent of Asians who were already council tenants in 1977 (Asian Census), 6.6 per cent of heads of household

(1978 Sample Survey) were on the council waiting list and would shortly be rehoused. For these people, their long-term aspirations would soon become reality. Without this anomaly, preferences are closely in line with present circumstances.

Active competition

As well as looking at how cultural preferences are reflected in the degree of satisfaction with existing accommodation and its location, it is also important to consider how they determine the extent to which Asians compete in the different areas of the housing market. Four areas merit analysis: local authority mortgages; improvement grants; improvement loans; and council housing.

In Blackburn there is a well-developed system of socially orientated mortgage provision. The criteria for gaining a mortgage are clear, and as in all such schemes àre designed to assist a quite narrowly defined group of individuals. Applicants are therefore likely to be drawn from one stratum of society and they will possess similar financial and socio-economic status. Analysis of Blackburn Borough data, however, shows that, despite this uniformity of applicants, there are discrepancies between white and Asian mortgage holders. Whites borrow larger sums of money to finance the purchase of more expensive property. Asian applicants, for example, borrowed an average of £2650 in 1977 to buy properties worth only £3080. This last figure is well below the ceiling price beyond which the Borough is not willing to lend (£5500). These data suggest that Asian applicants are not borrowing as much money as they might, and that they are choosing not to take advantage of the fact that the system allows the purchase of relatively more expensive properties. Restricted aspirations seem the most likely explanation for this. In addition, since the average price of Asian-owned property in Blackburn was only £4400 in 1978, this means that the majority of Asian owner-occupiers were eligible for local authority mortgages if they chose to apply for them. Only 34 actually did apply in that year.

The position with improvement grants also lends credence to the existence of differing value-systems in relation to housing. The Borough has canvassed grants energetically within the Asian population through the medium of staff who speak a variety of Asian languages, and through the extensive distribution of appropriate literature, also in a number of languages. Furthermore, within each Asian area affected by the legislation on housing improvement an information centre has been established and these are staffed by liaison officers who have an understanding of the background of the Asian population. Lastly, the Borough also employs a full-time Asian housing officer who can assist when needed. Despite all these measures, the Borough described the response of the Asian population to the opportunity of improvement as 'disappointing'.

To support the Grant Scheme, the Borough also provides improvement loans for those unable to finance the pòrtion of improvement costs in excess of the statutory grants, and those wishing to make non-statutory improvements. Again, however, the data suggest that Asian owner occupiers are either satisfied with their properties or that they have other calls upon their limited income which they consider to be more important. In the first eleven months of 1978 the Borough granted 45 loans, of which 23 went to Asians. The average loan requested by Asian applicants was, however, only 36 per cent of that requested by other applicants: the mean size of loan given to Asians was thus £1045 whilst that given to other applicants was £2883. It is possible to argue, though, that the difference in the size of loans resulted from Asian applicants already possessing better houses which therefore needed less improvement. This was not, however, true. In 1978, the mean values of properties owned by Asian applicants (as valued by the Borough) was £3538 compared to £4558 for the remainder of the applicants. Since Asians lived in cheaper housing one might have expected them to have borrowed *more* money for improvement rather than *less*.

The final area which provides an indication that many Asians in Blackburn do not share the same housing priorities as the white population is council housing. Chapter 2 contained a description of the positive manner in which the local authority has approached the problem of the decline of its owner-occupied housing stock, and how council housing has been allocated an increasingly important role in this task. The chapter also charted the growth of the public sector and the characteristics of its stock. This stock is allocated to applicants on a 'first come, first served' basis within four priority groups. These are (i) clearance victims, (ii) lodger families, (iii) priority tenants and owner occupiers. After the demands of these groups have been satisfied, houses are allocated to the fourth category, that of 'non-priority tenants'. There is no official residence qualification of the type which Rex and Moore (1967) criticized in Birmingham. Moreover the waiting list for accommodation is by no means large in relation to the stock, or building programme. In September 1982, for instance, the number of applicants on the waiting list was 104 clearance victims, 326 shared-facility tenants, and 2290 other priority tenants (cf. 1978 figures of 178, 314, and 1943 respectively). The tenants with shared facilities could consequently expect to receive housing within four months (depending upon the availability of vacancies on the estate of their choice) whilst other priority tenants would have to wait for around twelve months. The factors which were outlined in chapter 2 regarding the growth of the housing stock and the demographic characteristics of the town's population clearly coincide to produce a position where strain within the public sector is not unduly great. Perhaps because of this, the local authority appears to operate an applications procedure which maximizes choice and minimizes restriction. Although applicants are forced to re-apply each year (a process which prevents the accumulation of inactive applications) they are offered a choice of type, age, and location of property, subject of course to a vacancy becoming available. However, within reasonable limits those who are able, and willing, to wait for vacancies to occur on their first-choice estate are allowed to do so. Borough officials claim that they do not operate a forced policy of dispersal or segregation of Asian tenants and that they do not impose ceilings on the number of Asians who are offered tenancies on an estate. They argue that if concentration does occur, it is due to the preference of Asian applicants and existing white tenants who might choose to apply for a transfer. They are, however, aware of the fact that it is not helpful for an estate to gain a reputation as a 'ghetto' (as sometimes occurs because of sensationalist reporting in the local press) and they consequently adopt a flexible, but pragmatic, posture which is capable of adjusting to such circumstances as they arise. To an outsider, then, it seems as if the Blackburn system avoids many of the worst excesses perpetrated by local authorities against ethnic applicants. Only very prolonged contact with the system would indicate whether this is true but there are independent signs which suggest that it is.

Having described the public housing system in Blackburn, and its operation, the next step is to see how the Asian population has reacted to this. Analysis of unpublished Borough records (to which access was freely given) suggests that the Asian population effectively ignored the public housing sector in Blackburn prior to 1973/4. Reasons for this have been discussed in detail elsewhere (Robinson, 1980b) but, in brief, they include the prohibitive cost of council renting, the shortage of larger properties, and the absence of estates near existing cores of Asian settlement in the owner-occupied sector. One might perhaps add to these the additional reason that Asians were not waiting to be rehoused by the council during clearance programmes but were undertaking what has been termed 'bulldozer hopping': this is moving house just prior to demolition, to other areas which have been scheduled for clearance in the longer-term. One cannot, however, ascribe the very small number of council-housing allocations to Asians prior to 1973/4 to discriminatory practices on the part of the local authority. Some Asians did apply for housing and were given it. Moreover the Borough claims that very few, if any, Asians applied for council housing over and above those who ultimately received it. Since the majority of Asians in Blackburn live at high densities in two bedroomed terraces, the shortage of three and, particularly, four bedroomed council houses could not have been a major disincentive. One is therefore left with the explanation that few Asians had applied for council housing because few were willing to pay the extra cost involved and few wished to leave ethnic territory.

These influences seem to gain added support from the data concerning events since 1973/4. Since then, rent and rate rebate schemes have become both more widespread and more generous. Robinson (1980b) calculated, for example, that an average Asian family in Blackburn in 1978 might pay as little as 68 pence per week for a council house (after rebates). The economic incentive to move into the council sector cannot therefore be taken lightly. Secondly, in the mid-1970s, extensive areas of inner-ring terraced property were cleared and replaced by new, high amenity, council estates. These are consequently found in close proximity to existing areas of Asian settlement with their mosques, shops, services, and community networks. The move into council housing need not, therefore, be accompanied by isolation. Both these factors have been reflected in the number of Asian applicants for public housing, and their estate preferences. Although only two Asians had been allocated council houses in 1973, by 1976 annual allocations had risen to 22, by 1977 to 33, and by 1978 (the last year for which figures were available) to 56. Those who were

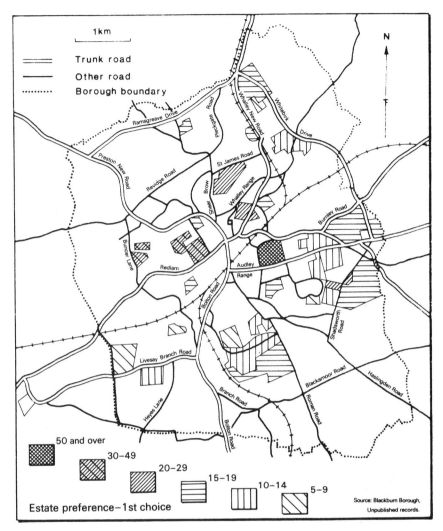

FIG 8.4: The pattern of Asian council estate preference, 1978

applying were largely naming the new inner-area estates as their first preference (see fig. 8.4 and Robinson, 1984b) and were sometimes turning down five or six offers of accommodation on other estates without even viewing them. As a result, they were spending twice as long on the waiting list as other applicants (11.5 months on average), but they were receiving houses on their chosen estates. The fact that these estates contain some of the best council houses in Blackburn means that, in marked contrast to most other towns which have been studied, the ethnic population in the council sector is very well housed. Only 0.8 per cent of Asian tenants live in pre-war properties whilst 69 per cent live in houses erected during the 1970s (cf. figs. 8.5 and 2.1). Although Asians occupy only 0.95 per cent of the town's publicly-owned housing, they occupy nearly 7 per cent of all four-bedroomed properties. In short, the Asian population has only moved into council housing when it became cheaper than owner occupation and when it did not require any loss of access to friends, relatives, and ethnic institutions. The fact that those who have become council tenants have also become better housed appears to be almost incidental.

This section has outlined the extent to which Asians participate in the housing schemes of white society and has discussed whether their motives for participation are similar to those of native whites. The persistent conclusion is that Asians in Blackburn do not participate in these schemes to the extent that their socio-economic position would lead one to expect.

A combination of the declared levels of satisfaction with existing

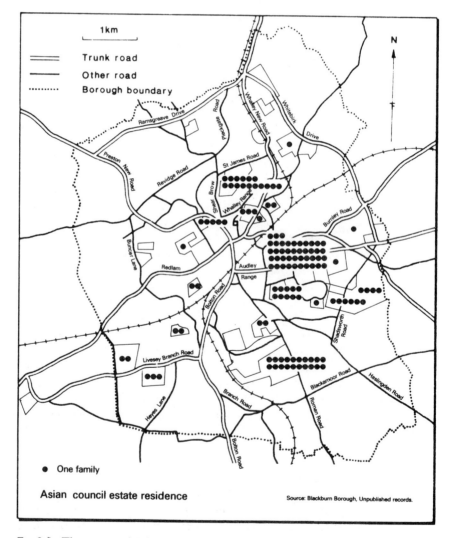

FIG 8.5: The pattern of Asian council estate residence, 1978

accommodation and an apparent unwillingness to spend a greater proportion of income on improving this, or on gaining upward residential mobility, intimates that the Asian population possesses different attitudes to housing as a commodity and a different scale of priorities, on which housing does not figure as highly.

Housing and the Asian Value System

It seems impossible to explain the residential preferences of the Asian population by recourse to either the British middle or working-class value-system. As a result, this section describes some of the cultural and economic tenets upon which the Asian attitude to housing in Britain is based. This may help to explain why Asians behave differently in the housing market from British whites of a similar economic status.

Perhaps the most important, and most obvious factor which differentiates the working-class white and Asian is their reliance upon different reference groups. Whilst sectors of the white population judge their own housing position in relation to the suburban ideal, this is not necessarily true of Asians. This argument has already been outlined in chapter 5, but it seems to have particular relevance for the Blackburn Asians. The fact that Asians are comparing their current housing amenities, size, value, and condition, not with a suburban ideal, but with their previous residence either in a village or small town in India, goes a long way towards explaining their apparent satisfaction with what are, by British standards, poor houses. Frequent visits, letters, and the arrival of new immigrants would maintain the sending society as the comparative yardstick. Low rates of car ownership, limited leisure time, and the absence of large-scale primary-group contact with whites might also reduce knowledge about the better housing standards enjoyed by the more affluent members of British society. In short, continued use of the standards of the sending society as a yardstick would mean that feelings of relative deprivation would be less likely to occur, and more limited efforts would be made to gain upward residential mobility in Britain.

Secondly, and to a certain extent a linked point, the myth of return seems to have relevance here. If a family is committed to raising its *izzet* in the village of origin, either by purchasing more land, a *pakka* house, or by giving a larger dowry, the acquisition of greater status in the eyes of white society through the procurement of larger and more expensive houses will inevitably be given a low priority. Only when those immediate targets which prompted migration have been achieved can thought be given to gaining status within the context of either the Asian or wider white society in Britain. The attitude to housing in Britain, at least in the short-term, will be that it should be the minimum possible drain on resources. This line of argument could clearly explain the timing of the move into council housing, the poor response to improvement schemes and the unwillingness of many Asians to commit themselves to borrowing larger sums of money. It also has a direct bearing upon the responses which Asians give to a question asking them what they would do with an additional £10 per week. 105 interviewees said they would either save or remit the money. Furthermore, the 1978 Sample Survey revealed that, on average, each household remits £19 per month to relatives abroad despite the fact that there are individuals who live in housing which is, to British eyes, substandard in almost every respect. It seems then, that the advancement of the joint family has been given higher priority than the immediate housing conditions of that part of it which lives in Blackburn. For an Asian in Blackburn, therefore, the characteristic of his accommodation which is of greatest interest is its cost. Many of the respondents in the 1978 Sample admitted that their main motive for coming to Blackburn was simply that houses were cheaper there than in other places in which they had lived. This certainly seems to be the case, for those 141 Asians from the 1978 Sample who declared the value of their monthly mortgage repayments, showed that these averaged £34.05. or 15.1 per cent of the mean household income. Remittances, by way of comparison, formed 8.4 per cent of income, whilst the unrebated rent for a council house might form only 12 per cent.

Lastly, chapter 5 discussed the way in which Asians in Britain avoid sources of cultural contamination in order that these do not prejudice their chance of returning 'home' later in life. It was argued also that many Asians would be unlikely to desire residential, or more intimate contact with British society and its people, since they could see a great deal to criticize and very little to praise. Seabrook (1971) provides a good example of this attitude in the comments of an Asian to whom he spoke in Blackburn:

Among the immigrants, if somebody in the family is sick, or if he has any problems, then the rest of the members of the family share the responsibilities and the pain. . .Now, in this country, I have found that the English people, especially the young generation, don't respect the older people any more. . .Nobody is ready to listen to them. Nobody helps them. The English people. . .do not always see that there are many things in the immigrants' lives that are better than the way the English people are living now. . .The immigrants' children do not say unkind and cruel things to their parents. But sometimes you hear Blackburn children say things to their mothers that you would not say to a dog. I see many social problems among the English that we do not have. . .There are the families who go to play the bingo and go to the pubs and so on. I've found that some of the families go to play bingo and leave their children alone in the houses. . .The parents need to be taught. They don't look after their children. Then they go to a pub and drink so much they lose their senses. . .and although English people marry after their own selection, are they happy? How many divorces take place in English families?. . .The custom of selecting a partner yourself is a good one, but they have taken advantage of the rule which their forefathers have made for them. . .Because the girls go for courting to the boys for selecting their future companions, there is no objection to sex before marriage. This is wrong in my eyes. But suppose you have the sexual intercourse before the marriage. . .Her only course of action is to find immigrant boys, or she becomes a prostitute. (Seabrook, 1971: 236-8.)

It seems that, if these views are widely held amongst the Asian population, and this appears to be the case in Blackburn, the whites are not the only group who think in terms of derogatory stereotypes. In the same way that these stereotypes discourage whites from seeking residential proximity to Asians it seems likely that many Asians do not desire primary-group contact with members of a society which they clearly believe has many faults. This is likely to support the initial spatial impact of chain migration through ports of entry, and will combine with the very obvious positive advantages of living in a community to produce a demand for some measure of self-segregation. The motives behind this need not be clearly thought out or particularly conscious. They do, however, explain the consistent theme of a dislike for moving outside ethnic territory.

Summary

This chapter began by looking at the attitudes which the indigenous white population of Blackburn hold towards what they regard as the homogeneous Asian population. These attitudes were characterized by the stereotype that Asians are a threat to many aspects of their livelihood and life-style. Moreover, Asians are not only seen as competitors but also as natural inferiors. These contentions were then supported by reference to the relative success enjoyed by ultra right-wing 'political' parties in Blackburn. In the light of this knowledge, data were then presented on whether such attitudes and prejudices were reflected in the power structure within Blackburn, and its spatial manifestations. Evidence proved that inter-ethnic segregation was substantial regardless of the scale or method of measurement, and that those areas of the town which could be classified as ethnic territory were markedly inferior on most, if not all, objective criteria. Moreover, Asian housing conditions did not result from conspicuous consumption in other directions, but were, at least in part, the product of exclusionary closure in the labour market. The Asian population has acted only as a replacement labour force and is still found in low-status, poorly-paid employment. This had clear implications for their position in the housing market. In addition, though, it appears that exclusionary closure exists independently within the housing market and acts to bar access to what whites regard as desirable housing or neighbourhoods. The conclusion is that white members of British society do, undoubtedly, practise widespread and effective exclusionary closure against Asians (as predicted by the housing class/underclass thesis and our own model) and that this potentially has far-reaching repercussions for social, residential, and spatial patterns.

The account then turned to look at whether the Asian population is engaged in usurpationary closure. The conclusions here were that, on the whole, they are not. They are clearly attracted by community living in ethnic space, and their attitude towards white members of British society seems unlikely to encourage a desire for residential or social mixing. Moreover, they express satisfaction with their current housing and the neighbourhoods in which they live. Finally Asians demonstrate only a limited desire to improve conditions unless it is accompanied by lower housing costs and ready access to existing ethnic space. It seems, then, that it is difficult to explain the residential behaviour of Asians in Blackburn within the context of British value-systems or

British housing classes and that it is therefore essential to investigate the Asian attitude to housing as a product of experience prior to migration. In this light, much of the behaviour then appears rational and understandable.

Overall then, the Asian population does not appear to be in direct competition with whites over housing or territory, although the latter seem unwilling to recognize this. As long as Asian attitudes to housing remain determined by the values of the sending society and the purpose of the migration, competition seems unlikely to escalate into conflict. If on the other hand the Asian population was to begin to judge its housing conditions by standards prevalent in British society, it would surely compete more fiercely for access to resources. In doing so, it would encounter the full force of exclusionary closure, and its general disposition towards British society and its indigenous peoples would surely change profoundly. Moreover, active usurpationary closure by Asians would serve to validate many of the stereotypes held by whites and would be likely to intensify the strength of feeling against them. It seems then that, at the present time in Blackburn, a 'phoney war' exists. Full scale conflict can only be averted if this temporary truce is used to dismantle exclusionary barriers and change the attitudes of the indigenous population.

CHAPTER 9

East African and South Asians: Contrasts and Similarities

Previous chapters have demonstrated how South Asians and East African Asians have different pre-migration experiences, and contrasting reasons for being in Britain. In addition, it is likely that the two groups differ in the extent to which they see residence in Britain as a permanent phenomenon. As a result of these factors, it was suggested in the model that the South Asian minority could be characterized as being encapsulated whilst the East Africans might be thought of as a marginal group, but one engaged in usurpationary closure. This chapter investigates the empirical evidence for this notion and does so under five major headings: financial considerations; patterns of residence; employment; education; and perceptions of British society. These headings have been selected from the range of alternatives for a number of reasons. First, the scale of priorities which controls expenditure also determines the strength of an individual in many of the markets he is forced to enter; second, residence not only reflects social distance and acts as a symbol of status, but also influences access to services, and therefore life-chances; third, the workplace is an important arena for the formation of primary and secondary group relations which might either emphasize (Hechter, 1978) or dilute ethnic solidarity, and employment is also the chief means of gaining resources (wages); fourth, the school in the educational system forms not only a method of social selection for adult life, but also the crucible in which adult ideas are forged and hardened. The school also provides the key focus around which the primary-group contacts of children are arranged; and lastly the attitudes of individuals naturally colour their perceptions of British society and its constituent individuals. In this way, perception can become self-fulfilling.

Financial Considerations

The discussion so far has stressed the integral part which remittances play in the myth of return and in providing a motive for migration to Britain. In chapter 8 it was suggested that 'Asians' have different financial priorities from those of local whites and that the former place saving and remittances above immediate expenditure on improving housing standards. Chapter 5 stressed that remittances have not only a financial purpose but also a cultural and emotional one since they act as an important symbol of commitment to return migration and hence to the traditional ways of the sending society. It is hardly surprising, therefore, to discover the size and significance of cash remittances from Britain to India and Pakistan. At a national level, data on remittances are rarely made available to non-government users, but several sources have provided unpublished data specifically for this study. However, before these data are presented, it is pertinent to note that it is widely acknowledged that declared cash remittances are but a fraction of the true total. Other channels include 'black' money, insurance policies, smuggled cash, and consumer durables. In Pakistan there is also the option of importing cars and vans which can then be resold at a profit of some £500 per transaction. Some idea of the scale of these transfers of resources can be gathered from the fact that two days of observation in Mirpur (Pakistan) revealed some 62 vehicles still travelling on British number plates. Although opportunities with vehicles do not exist to the same extent in India, possession of foreign currency does allow access to what are known as 'export quality' goods such as 'Baja' scooters or Ford tractors. Foreign currency also circumvents the extensive waiting lists for these items and negates the need to pay premiums for early delivery. When there are waiting lists of up to five years for a (Fiat) Premier Padmini car this is no slight consideration.

Despite the fact that the size and importance of remittances to India and Pakistan are thus considerably understated by official data, these themselves make impressive reading. Table 9.1 provides a simple statement of the amount of money remitted by all overseas Pakistanis during the years 1970–84. They have been converted approximately from dollars into sterling to allow comparability, but their major interest lies in the phenomenal increase which has occurred in the value of remittances over this period. They have increased 56 fold, and whilst a large proportion of this can be attributed to the recent movement to the United States and, more importantly, to the Middle East (see Shah, 1983 and Robinson, 1986), the increase does also reflect a rise in remittances from established communities. Table 9.2 demonstrates how the British Pakistanis have contributed to this overall growth, although the value of their remittances 'only' trebled during the ten-year period 1973–83. Because of the new wave of emigration to other destinations, the percentage which remittances from Britain form of the total has, however, fallen from 31.9 per cent in 1973 to 4.9 per cent in 1983.

The data on remittances from Indians overseas are both less comprehensive and harder to interpret. The official government stance on data relating to private remittances is that they do not exist. Despite this, there have been various attempts to estimate remittances from Indians overseas. Helweg (1983), for example, suggests that gross receipts in 1980 were as high as £2628 million, although his figure includes not only personal remittances but all gross non-export receipts from shipping, insurance, tourism, and dividends. Informal contact with officials at the Reserve Bank of India in Bombay revealed that data actually do exist on private remittances but that these are not normally released to individuals other than bank employees and government researchers. These data are presented in table 9.3 for the years 1960–79, based on a constant exchange rate of 17 rupees to the pound sterling. Whilst officials at the Reserve Bank admitted that they did have data relating to the proportion of these totals sent by Indians in different overseas countries, they declined to make these available since they recognized that they were inaccurate. A study commissioned by the Bombay Chamber of Commerce (1978), however, found that

33 per cent of all remittances sent to Gujurat came from the Middle East whilst a further 26 per cent came from the UK. If the ratio between recorded and actual remittances to India for 1973 (which is known) held good in 1978–9, it is likely that total personal remittances to India in that year were approximately £625 million, of which perhaps £150 million were sourced from the UK. Clearly then, personal remittances by Indians and Pakistanis in Britain represent sizeable flows of capital, even in an international context.

Field research soon demonstrates that official data, whilst valuable, bear little relation to the total economic value of having family members working in Britain. Government officials in Delhi revealed that research had demonstrated that the average family in Gujurat received £64 per month in remittances. The author's own field research in the Punjab, Gujurat, and Mirpur areas revealed average cash remittances which ranged from £6 per household per month through to £75 per household per month in 1982.

However, in order to evaluate the impact of remittances on village life and facilities, and upon the disposable income of the remitter in Britain, it is probably more valuable to consider specific cases at a local level. Field research in the Jhelum district of Pakistan during 1982 revealed two especially interesting cases. The village of Bishandaur (pop. 2000) has approximately 100 people living and working abroad, of which several reside in Oxford. Of these, Sher Afzal Khan is the most prolific remitter and his own money, and that raised from others, succeeded in 1966 in paying for one of the village's roads to be surfaced. Since then he has organized campaigns to build a wall around the village graveyard, to help extend and equip a medical dispensary in the village, and to get the school upgraded to high school standard. His latest effort is an attempt to build a new bridge across the nearby railway line but this appeared not to have been completed by the time of my own visit. The second case involved a man from a nearby village who is now employed as a taxi driver in Milton Keynes. He had lived in Britain for sixteen years and was nine years of age on his arrival here. He told me that over a period of two years he had saved £900 in Britain and that, with this, he had arranged for a well to be sunk in his old village. The walls of the well had been lined with bricks and cement to provide a permanent facility. His efforts had saved the villagers a one and a half mile walk which they had previously had to make to get water. He had also just spent £300 renovating and re-tiling the village mosque and was now saving to pay £250 to have a local road resurfaced. Whilst the philanthropy of these two cases might be unusual I came across other cases where emigrants had made similar efforts to save money, but for the benefit of their own families rather than for the entire village.

It is clear then, that at both a national and local level, remittances are a significant element in the economies of the sending society and village (see Robinson, 1985c for further details). However the major concern here is their impact on the levels and patterns of expenditure of the migrants whilst they are in Britain. The 1978 Sample Survey looked into this. Of those 254 South Asians who were willing to declare whether they sent remittances or not, 169 admitted that they did. In other words 66.5 per cent of South Asians still sent remittances despite the fact that they had, on average, been in Britain nearly nineteen years. Of these 169, 156 were willing to specify the amount which they remitted per month. This proved to be an average of £19.34 (at 1978 values). Taken together with a separate question on average monthly household income (including overtime), this indicates that remittances formed approximately 9 per cent of total income, or 60 per cent of the amount which an average household spent on accommodation in Britain. It also seems as if these remittances are sent, not because the family in Britain has reached a comfortable position and can easily afford to disburse £19 per month, but in spite of the fact that the remitters are poor and can ill-afford it. When asked to what use they would put an extra £10 per week, almost 54 per cent of South Asians said that the money would have to be used simply to survive (i.e. pay household bills, or buy clothes). This contrasts with only 0.8 per cent who would invest in household durables, 1.9 per cent who would simply spend and enjoy the money, 1.2 per cent who would use it on entertainment, 2.3 per cent who would buy a car and 2.7 per cent who would improve their life-style. It seems that, for South Asians at least, economic survival is still a struggle, but that despite this, remittances are still thought to be a major priority. This is underlined by the fact that 29.5 per cent of respondents would save the extra £10 per week for their return rather than use it now on their immediate needs. A further indication that the South Asians are not remitting money only *after* they have indulged in conspicuous consumption in Britain can be gained by considering their rates of ownership of consumer durables. 38.5 per cent of South Asian respondents in the 1978 Sample owned either a manual or automatic washing machine, 26 per cent had a telephone, 59 per cent had some form of cassette recorder or stereo, 32 per cent owned a colour television and 37 per cent a car. Comparison of these rates with either national or local averages indicated that the South Asian population is indeed deprived by British standards. The results of the 1981 Census, for instance, show that 50 per cent of all households in Blackburn (including Asian households) own a car.

It would appear, even from this brief survey of national and local data, that immediate remittances and saving for return migration remove a significant proportion of household income for South Asian families. This saving takes place despite the relative poverty of the families concerned and it received a high priority within household budgets. The removal of at least 10 per cent of total household income (a conservative estimate) cannot but disadvantage South Asians in the struggle for resources, unless their incomes are considerably higher than their white working-class counterparts. If this is not the case, then South Asians cannot hope to have the same choice of housing or the same level of material comfort. Moreover, in areas where finance can secure access to preferential services (such as private medicine or education) they are again electing to forego these opportunities. They are thus trading off the immediate needs of the extended family for social mobility or social stability, against the long-term prospects for their own nuclear family in Britain. This, in itself, is a comment upon the level of commitment to the sending society.

The situation in relation to East Africans and their financial priorities is somewhat complicated by the presence of dependants in India who are still awaiting entry under the quota system. In many cases, these people expect, and are sent, remittances to maintain them prior to their re-emigration. Despite this complication, it is clear from the Blackburn data that far fewer East Africans remit. Of those 92 individuals who responded to the question on remittances, only 35.8 per cent did still send money either to India or East Africa. This is slightly more than half the comparable figure for South Asians, and one must bear in mind that the East Africans have been in Britain for considerably less time and, therefore, might reasonably be expected to have stronger links with their homelands. Those East Africans who do remit, however, send larger amounts of money, and this underlines the expectation of a higher living standard possessed by many East Africans, based upon their experience prior to expulsion. The average remitted by those East Africans in the 1978 Sample proved to be £25.03 per month (at 1978 values).

The fact that East Africans are more reluctant remitters begs the issue of which financial priority is given greater emphasis instead by this group. The 1978 Survey question on ownership of consumer durables provides some answers. Forty-seven per cent of East Africans owned a colour television against 32 per cent of South Asians; 34 per cent of the former rented telephones against 26 per cent of the latter; and 75 per cent of the former versus 59 per cent of the latter owned a stereo system. Similar, but smaller, differentials existed for the possession of refrigerators and washing machines. A second indication of the priorities of East Africans can be found in the results of the 1978 question which enquired about the use to which an extra £10 per week would be put. Almost twice as many East Africans as South Asians mentioned improving the quality of their lives. Specific items were better housing (74 per cent), a higher standard of living (4.2 per cent), more entertainment (4.2 per cent), family holidays (1.1 per cent), and immediate consumption of items such as household goods (4.3 per cent). It appears from answers such as these that East Africans, in Blackburn at least, are striving to improve their life-style and accumulate material goods in a way that the South Asians are not. This may well represent an attempt by the East Africans to re-create the middle class life-style and social position which they enjoyed before their expulsion.

However a tendency towards conspicuous consumption is not the only element in the financial strategy of the East African group. Given their mercantile heritage and merchant ideology (Tambs-Lyche, 1980) it is not surprising that saving for later investment also figures strongly. East Africans are already more heavily represented in the self-employed category in Blackburn than their South Asian counterparts, despite the fact that they have been in Britain less time. However, this return to self-employment is not gained cheaply and, as both Mullins (1980) and Jones (1982) have demonstrated, even ownership of a corner shop in Croydon or Leicester can require £40,000 in capital. Clearly the need to accumulate such sums is likely to restrain conspicuous consumption in the short- and medium-term despite the group's higher expectations and aspirations. For individuals such as Messrs Shamji (Batchelor, 1984) or K.D. Patel (Smith, 1984), however, these few years of austerity have proved to be a sound investment and they are now reaping the rewards for this in their life-styles.

Housing

The fact that both the South Asians and the East African Asians have such clear financial priorities, particularly with regard to saving, suggests that housing may well be seen as a low priority and consequently be allocated few resources. It might also be reasonable to expect that East Africans should evidence higher levels of dissatisfaction with their accommodation than their South Asian counterparts, given their middle-class status in East Africa. And one might also predict that East Africans would be both better equipped and more willing to compete with local whites for housing resources.

In practice, the two groups live in property of a very similar value (average cost of £4454 for South Asians and £4397 for East Africans) and of comparable quality, and this despite the East Africans having lived in Britain for an average of seven and a third years less. They have succeeded in achieving this parity by allocating a greater proportion of their earnings to the provision of accommodation: average monthly repayments for East Africans were £40.63, or 17 per cent of total household income, whilst those for South Asians were £32.71 or 15 per cent of income. This indicates that not only have East Africans immediately demanded the same quality of housing that South Asians possess (despite their more recent arrival) but that they have been willing to allocate a greater proportion of their resources to achieve this goal. This hints at the greater aspirations of the East Africans and suggests that, in time, this group will be found in better housing than that of the South Asians.

Moreover, East Africans have achieved this parity in an accelerated manner by using what Rex and Moore (1967) termed 'legitimate' methods more frequently than have South Asians: 37 per cent of East Africans used non-ethnic means of locating their present house against 29 per cent of South Asians; 42 per cent more South Asians used subsidized local-authority mortgages; four times as many South Asians used loans from finance companies; but 14 per cent more East Africans used building-society mortgages or cash. These results are also supported by a comparison with those of the CRC (1977) for Bradford (Pakistanis) and Leicester (East African Asians).

Although the housing standards of the two groups appear to be broadly similar, the contrasting value-systems and expectations of their members produce differing levels of satisfaction: one-third fewer East African owner occupiers than South Asians declared themselves to be very satisfied with their existing homes; 40 per cent more East Africans expressed either uncertainty or dissatisfaction about their housing; four times more South Asians than East Africans thought their houses were too large; and half the East African respondents mentioned at least one item that they disliked about their house or neighbourhood, against 38 per cent of South Asians. Additionally, when asked whether they would prefer to improve their current house or move to a different one, 31 per cent of East Africans would opt to move whilst the comparable figure for South Asians was 22 per cent. Lastly, whereas almost 26 per cent of South Asians said that they had chosen to live in their present neighbourhoods because of the positive aspects of its physical appearance and houses, the same could be said for only 17 per cent of East Africans. Whilst many of these discrepancies in levels of expressed satisfaction appear small in their own right, when taken together they are strongly suggestive of the fact that East Africans are less satisfied than South Asians with comparable housing. Clearly their pre-migration experience leads them to expect and demand more. This expectation is also manifested in levels of self-esteem. When asked of which social class they regarded themselves as members, 32 per cent of East Africans said 'middle class' whilst the remaining 68 per cent thought of themselves as 'working class.' The same question elicited a different response from South Asians. Of these, less than 20 per cent said 'middle class', and over 80 per cent said 'working class' (see Jenkins, 1971 for corroboration of this point). The East Africans thus have a self-image which is more likely to involve them in the acquisition of symbols of their status, and in the preservation of social distance between themselves and other immigrant groups who are perceived as status threats. In this sense, the attitude of East Africans towards Indians, and more particularly Pakistanis, is somewhat similar to that of the white working class. It is not too suprising, therefore, to see the East Africans taking active steps to exclude other Asians from their residential neighbourhoods (see Phillips, 1981 for a good description of this process at work in Leicester).In Blackburn, though, the East Africans have yet to reach a position where they control the institutions which can operationalize such desires for exclusionary closure. The attitudes which underlie these desires, however, *are* present. During the 1978 Survey, respondents were asked to place a variety of ethnic groups in rank order of preference. Amongst the East Africans, only 1.9 per cent named the Pakistanis as their first-choice group. At the other extreme, a full quarter of East Africans named the Pakistanis as their least preferred group. The reasons behind such preferences became clear during fieldwork in 1982 in the Kheda district of Gujrat State. There, the settlement of Villabh Vidyanagar was built specifically to house returning Patidars who were leaving, or being expelled, from East Africa. East African returnees there were unanimous in their condemnation of Pakistanis. They argued that it was this group who generated feelings of racism amongst the white population in Britain because the Pakistanis were uneducated factory workers who had militant tendencies, were often in the vanguard of strikes, exploited the immigration rules, practised polygamy, and even married their own sisters. They also criticized the Pakistanis for their limited attempts to become fluent in spoken English. Only later did the East African interviewees express any opinion on India and the average Indian and these were initially expressed only in a muted way. Only as the discussion developed would interviewees give vent to their true emotions. They described how they were regarded as foreigners by Indians and how their educational achievements and higher standard of living gave rise to feelings of envy and resentment. These feelings occurred not only at an inter-personal level but also in dealing with government institutions. They were particularly critical of the attempts to integrate the scheduled castes through positive discrimination and quotas, and ruefully described the injustice and mediocrity which resulted. They also despised the corruption and nepotism which they considered existed at all levels within Indian society. They concluded, and summarized their feeling, by describing Indians as *anamat* which they translated as meaning 'backward class'.

Although these attitudes probably prevailed more amongst the educated crown pensioners in Villabh Vidyanagar than they do amongst many East Africans in Britain, clear evidence of social distancing can be derived from the Blackburn data on residential separation. Again, this can be measured in a number of ways and at different scales. Table 9.4, for instance, provides indices of dissimilarity for households at enumeration district level in 1977. They reveal that, whilst the degree of residential intermixing between Indians and East Africans is considerably more than between either of these two groups and the Pakistanis, it is nevertheless by no means complete. Polarization does exist between members of these two groups despite their common heritage, religion, and, in many cases, language. As was noted above, the Pakistanis are the most polarized of the groups, not only with regard to other Asian groups but with regard to the white population. Either the Pakistanis choose to enforce a higher degree of self-isolation or they are the group from which others choose to maintain the greatest social distance. Table 9.5 can be interpreted in a very similar manner: the values here are for individuals at ward level and rely upon categorizations by birthplace, not ethnicity. On this criterion, and at this scale, the Pakistanis again seem to be the group from which others are keenest to separate themselves. This even goes so far as to produce a situation where East Africans are actually less residentially polarized from the white population than they are from the Pakistanis. Levels of Indian–East African separation are again low but persistent. Table 9.6 develops the analysis by considering group isolation, measured by P* at ward level in 1977 for adult Asians and their children of British and non-British birth. P* shows that all groups are considerably more isolated in residential terms than one might expect, given the proportion of the total population which they form. This level of isolation, over and above the expected, ranges between 2.77 times the expected for the Indians to 4.33 times for the Pakistanis. This suggests that, whilst the East Africans are the least isolated in absolute terms, the Indians are the least isolated in relative terms.

The measures described above concern themselves with residential intermixing, not with territoriality. They show that the East Africans and Indians vie for the position as the least polarized group but that the Pakistanis are consistently the most segregated. Figures 9.1, 9.2, and 9.3, however, concentrate upon the distribution of ethnic space. They do not aim to provide accurate maps of the distribution of members of the three groups but rather to illustrate which areas are important centres of their settlement. Location quotients were calculated for each group in each ED in which they were represented. This reveals a major Indian core area to the south east of the town centre along Audley Range and to the north east; Pakistani areas to the south, north, and east; and East

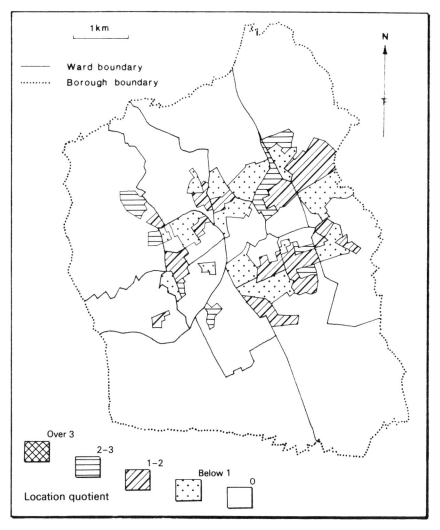

FIG 9.1: The distribution of ethnic Indians in Blackburn, 1977

African core areas to the west and north east. One further point which arises from these maps is the presence of isolated East African EDs either outside the main areas of ethnic settlement or within what could be thought of as another group's territory. The explanation of these outliers requires a discussion of the role of council housing.

The recent movement of Asians into council housing was commented on in chapter 8, but further analysis of the records of Blackburn's housing department shows this not to have occurred to the same extent for every sub-group within the total Asian population. In 1973–4, when there were only seven council houses allocated to Asians in Blackburn, five applicants were East African Asians (Ugandan Asians from the Greenham Common Transit Camp). This trend has continued, and by 1979 East African Asians formed 43 per cent of all traceable Asian council tenants although, in the town as a whole, they formed only about 15 per cent of Asian individuals on the criterion of birthplace, and only 22 per cent of Asian households on the criterion of ethnicity. This more extensive use of public housing by East Africans cannot however be explained by force of circumstance. The 1978 Survey showed that significantly more East Africans viewed council renting as their 'ideal' tenure, and that, in addition, some 13.3 per cent of East African heads of household were already on council waiting lists. This figure compares with only 4.4 per cent of South Asians. Clearly then, East Africans have been more alive to the economic advantages of public housing and more willing to take advantage of this. Again this provides evidence of greater competitiveness and a willingness to engage in usurpationary

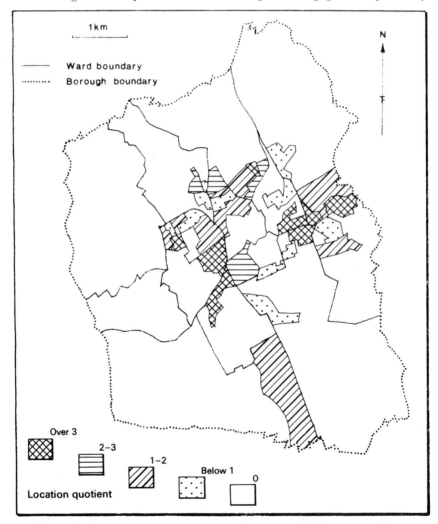

FIG 9.2: The distribution of ethnic Pakistanis in Blackburn, 1977

closure. In return, the group enjoys centrally-located, modern and attractive housing at a fraction of the cost of owner occupation. In short, they have already taken the first step towards ascendancy over South Asians in the housing market rather than mere comparability. This is not peculiar to Blackburn (Bristow and Adams, 1977; Phillips, 1981).

This section has described how the East Africans are both more competitive and more successful in the housing market than are South Asians. Despite the handicap of arriving in Britain later, East African owner occupiers have already achieved similar housing standards to South Asians and have additionally been in the vanguard of the movement into public housing. To achieve their success, they more frequently use methods defined by white society as

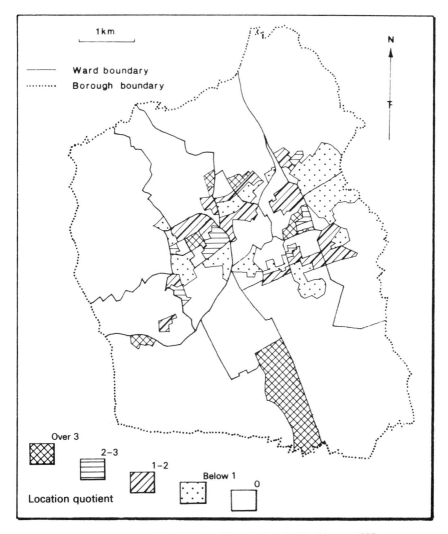

FIG 9.3: The distribution of ethnic East African Asians in Blackburn, 1977

legitimate, and appear also to judge their housing by standards more akin to local norms than to Indian and Pakistani ones. Finally, a strong financial motive underlies many of the strategies adopted by East Africans. South Asians, on the other hand, are also economically motivated but they more often lack the willingness or ability to pursue this goal when it involves direct competition with the local white population. Moreover their perception of housing standards in Britain is more frequently determined by pre-migration experience, and because of this, standards are considered satisfactory. Finally the improvement of housing conditions appears to be given a low priority.

Employment

Chapters 3 and 8 developed the notion that Asians have been used as a replacement labour force concentrated into those industrial sectors unable to attract sufficient indigenous labour and it was also shown that this appears too to have been the case in Blackburn. Again, however, further analysis shows that differences do exist between East Africans and South Asians. Table 9.7 demonstrates that the East Africans rely much less upon the textile industry for employment than do Indians and Pakistanis. The former, consequently, enjoy a more balanced and less vulnerable employment structure. These differing levels of diversification can readily be quantified by using the Gibbs-Martin Index (see Gibbs and Martin, 1962). This ranges from zero (complete concentration in one industry) to one (maximum diversification) although Clemente and Sturgis (1971) have shown that the upper limit of the index varies according to the number of categories involved. Using Department of Employment (1979) data and the Standard Industrial Categories, it is also possible to calculate a control figure for the North West Region as a whole. In 1977, then, the Gibbs–Martin Index for the regional work-force proved to be 0.852, that for the the East African population of Blackburn, 0.663, and that for South Asians in the town, 0.596.

It is also important to note that the degree of diversification in the employment structure of the East African population is not simply brought about by a more even spread throughout the town's *declining* industries. As was pointed out in chapter 2, Blackburn now has a thriving and expanding engineering and electronics industry. That a secondary concentration of East Africans exists in these buoyant sectors bodes well for the future since many employers are actively engaged in advanced electronics, a sector which the OECD considers will form the pole 'around which the productive structures of advanced industrial societies will be reorganised' (*Economist*, 1981). The East Africans would thus seem to be more sheltered from structural or cyclical unemployment and under-employment, and are well-placed to acquire the skills which are in demand within the labour market.

An additional facet of their more diversified employment structure, is the East Africans' greater presence within the distributive trades. The discrepancy between the East African and South Asian representation in this category, is, in practice, more marked than the data indicate. Invariably, the interviewers in the 1977 Census would report that East African heads of household gave their own occupation as a shopkeeper but said that their wives or daughters did not work despite the fact that they were known to help in the family shop. Conversely, South Asian retailers frequently gave the employment of *all* their family members as shopkeepers. This inevitably influences the results which appear in Table 9.7. However, it still remains uncertain as to whether retailing will act as a route to economic mobility (Aldrich, 1980), although it does undoubtedly provide a measure of economic independence and it does represent a means of avoiding the many discriminatory practices which bar the advancement of ethnic workers. At the very least, then, the East African movement into retailing represents the re-emergence of entrepreneurial talent and a manifestation of the desire for economic and social mobility.

A further indication of the differing orientation of East Africans and South Asians comes from an analysis of the ethnic composition of the work-force in plants in which these groups have sought employment. Respondents in the 1978 Sample Survey were asked to give the name and location of their current employer. This allows a consideration of whether Asians are concentrated into a small number of plants, or whether they are more widely dispersed. The analysis suggested a high degree of concentration in a relatively small number of plants. Those respondents who were employees listed only seventy-four different places of work although again there were significant differences between the South Asian and East African groups. For the former, almost 60 per cent of respondents were found in only eight plants and the largest of these alone employed over 10 per cent of the group. If the 1978 Sample is representative of the town's entire Asian population, nearly 76 per cent of South Asians are employed in plants where there are at least another ten Asian workers. The converse is that only 19 per cent work in plants where there are less then ten Asian employees (the remaining 5 per cent being self-employed). This suggests that encapsulation is present not only on the criterion of industrial sectors but also on a plant-by-plant basis within these sectors. Further analysis of data derived from the 1977 Asian Census and 1978 Survey, reveals whether Indians and Pakistanis continue to maintain social distance in the workplace as they appear to do in the residential context. In total, 31 major employers were included in the analysis, and these accounted for 75 per cent of the economically active respondents in the 1978 Sample. Figure 9.4 demonstrates the results of this analysis by illustrating the dominant regional-linguistic group at each of these major employers, as defined by the location quotient. In the case of both the Punjabis and Gujuratis the sample size was 81 persons. The map reveals the ways in which divisions within the Asian population influence the place of work such that, in general terms, the south and south east of the town's hinterland are dominated by Punjabis whilst the north, east and south west are dominated by Gujuratis. This pattern reflects the location of ethnic core areas (see figures 9.1, 9.2, and 9.3) as well as the orientation of major routeways and public transport. For example the workplaces in the Punjabi core areas immediately to the south of the town centre are also Punjabi dominated, as are plants sited in towns along the main road to the south (Darwen and Bolton).

Conversely, the Gujuratis predominate in workplaces within their north eastern core area as well as in towns located within the sector (Great Harwood and Oswaldtwistle). One should not overstress this spatial separation, however, as the majority of workplaces appear to afford employment to members of both groups. Having said that, the separation of the two major groups at their place of work *is* important, and in Blackburn it gives rise to segregation of 51 per cent as measured by the ID (calculated only for the thirty largest employers). Moreover, this situation is not unique to Blackburn in view of the relatively widespread practice of chain recruitment into ethnic work gangs. A frequent corollary of such encapsulation is minimal interaction between ethnic workers and their white counterparts. A second corollary is that South Asian

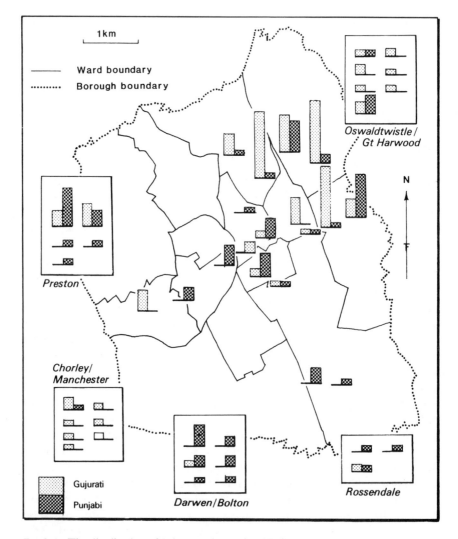

FIG 9.4: The distribution of Asian employees in 1978 by regional-linguistic group

workers are forced to travel longer distances in order to ensure that they work at a plant where the other employees are from the same regional-linguistic group. In addition, reliance upon go-betweens (Wright, 1964) or brokers to arrange employment also enforces spatial mobility.

Another facet of South Asian encapsulation in the workplace which is frequently mentioned in the literature, is that of shift work. Almost one-third of South Asian respondents in the 1978 Survey admitted to working on a night shift where ethnic work gangs are again common. In many cases, South Asians chose to work these shifts to maximize their income, regardless of the effect that this was likely to have upon their family and social lives. On average, the 1978 respondents of South Asian origin worked 41 hours per week, which represents a weekly average of 3.5 hours overtime. Despite this, their average weekly household income proved to be only £51.93 when calculated for all households and £59.47 when calculated only for those households where the head was in full-time employment.

In contrast, the East African respondents exhibited different characteristics. They were found to a greater extent in the smaller plants where few other Asians worked. In the 1978 Sample, few East Africans worked in plants where there were more than two other Asian respondents, and those who did, worked in skilled jobs such as electrical engineering or aerospace engineering where work gangs are less common. Their greater willingness to work for employers who did not have a reputation for hiring ethnic labour also meant that the length of their journey to work was shortened. Almost one-half of East African respondents in the 1978 Survey had a journey-to-work time of less than ten minutes. This compared with 32 per cent of South Asians. The East Africans also differed in the extent to which they worked shifts. Only 7.9 per cent of East African respondents said they worked on the night shift and a full 44 per cent declared that they worked only on a day shift. This last figure can be contrasted with the 19 per cent of South Asians who admitted similar working arrangements. However, the East African sample did, on average, work more hours per week, although any such conclusion has to be tempered by the inclusion of a number of shopkeepers who said they worked 56 and 60 hour weeks. Even so, the average East African worked for thirty minutes more per week that his South Asian counterparts although the modal value proved to be 40 hours per week. This extra overtime would be unlikely to account for the substantially higher household income of East African families who earned an average of £58.80 per week (all households) or £66.14 per week (only households where the head is in full-time employment).

Another issue on which the two Asian populations diverged was that of female employment. Of the adult South Asian women interviewed in 1977, only 8.9 per cent were currently in employment, whilst a further 0.8 per cent were unemployed and 7.6 per cent were unavailable because of pregnancy or the need to look after small children. In total then, 17.3 per cent of the female labour force was available for employment either in the short-term or in the near future. Of the East African women respondents in the 1977 Asian Census, 11 per cent were currently working and an additional 15 per cent would enter the labour market when circumstances permitted. The 1978 Survey shed extra light on this issue, since female interviewers questioned women respondents about the exact reason why they were not working. Of the 85 East African interviewees, 16.5 per cent were presently working, 17.6 per cent were not working at the moment but had in the past and could in the future, 31 per cent had never worked but would be willing to if conditions changed, and 35.3 per cent had never worked and either did not wish to or were not allowed to. Women in this latter category were barred from employment either by the wishes of their husbands, religious beliefs, health, or old age (in descending order of importance). Of the 231 South Asian women respondents, 9.1 per cent were in employment, 6.9 per cent were not working but had done so in the past and could in the future, 21 per cent had never worked but could if circumstances changed, and a full 63 per cent seemed unlikely ever to work. The average working week for both groups was approximately 39 hours.

The data which have been presented on patterns of employment lend considerable support to the conclusions derived from other information. The East Africans are employed in a wider range of industries and this range includes several expanding and successful sectors. In addition they have used their entrepreneurial skill to re-establish themselves in the distributive trade where they are in direct competition with white traders. They work in plants which have few other ethnic employees and they are not often employed on night shifts. They have a shorter journey-to-work than South Asians and earn more money. Finally their wives and daughters have already entered the labour market to a (marginally) greater extent and are willing to do so to an even greater degree if conditions allow or require it. The South Asians have remained encapsulated not only in their range of employment but also in the plants in which they work and in work gangs within these plants. Despite these objective findings, the subjective interpretation of these does vary. Almost twice as many East African Asians as South Asians (13.2 per cent *vis à vis* 6.8 per cent) declared themselves dissatisfied with their jobs. Furthermore, the level of dissatisfaction was higher in the larger plants where other Asians were employed than it was in smaller, non-ethnic workplaces. 36.4 per cent of those East Africans who worked in the latter were very satisfied against 5.4 per cent of those who worked in the former. The reverse proved to be the case amongst South Asians. 9.5 per cent of those who worked in encapsulated conditions were very satisfied compared with 5.4 per cent of those who worked outside the major employers. This subjective dissatisfaction with the conditions of employment amongst East Africans, even though their conditions were objectively better, again hints at the differing aspirations and motivations of the two groups. This interpretation is further strengthened when one considers the social class structure of the South and East African Asian populations (see table 9.8) as derived from the full 1977 Asian Census. Although differences are still small, the East Africans already appear to have achieved a greater representation in the higher social classes. Clearly though, their present position still does not match their self-image and they will therefore continue to strive for greater social mobility, social status, and access to resources.

Education

The data which exist concerning differing attitudes towards education within the Asian population and the differing behaviour of each of the groups' children are neither conclusive nor complete. Seen in isolation, many of the discrepancies appear to be trivial and of dubious statistical significance. Nevertheless, they do seem to be consistent with what has already been argued and they are therefore provided as being illustrative rather than conclusive. The data do, however, provide valuable confirmation for the notion of South Asian encapsulation in an area where this phenomenon is not often studied.

Many local authorities organize school attendance on the basis of spatially contiguous catchment areas (Williamson and Byrne, 1979). This ensures that those Asian children, whether Indian, Pakistani, or East African, who are resident in the inner-city areas attend schools where many, if not actually a majority, of their classmates are also Asian. In this way encapsulatory tendencies may be enhanced to the detriment of the children's spoken English. This consequently has to be learned rather than adopted through casual everyday usage. The predominant mode of communication in schools in which Asians are concentrated is naturally Punjabi, Gujurati, or Urdu, and English is used only in formal conversation in the classroom. Outside this, in the home, in the street of residence, in the cinema, and in the playground, the child speaks his inherited mother tongue (see CRC, 1976 for further data). The organization of school catchment areas can thus itself become a force towards encapsulation rather than integration. However, the younger child not only has to accommodate the use of the mother tongue as a passive, inherited phenomenon, but he or she also has to sustain the active steps taken by parents to develop its use through enforced attendance at mosque schools (see Nagra, 1982 for a discussion of the activities of Asian supplementary schools). Within these schools formal classes are organized in religious education and, more importantly, instruction is given in both spoken and written Asian languages. The importance which parents attach to such education is revealed by the answers which respondents in the 1978 Survey gave to questions asking whether they considered it desirable that children should be able to speak and write their inherited mother tongue. Around 90 per cent of both South Asians and East Africans thought it desirable that children should possess both abilities, and nearly three-quarters went on to add that the child's parental language should be formally taught in state schools. In each of these cases the differences in response from South Asians and East Africans were not great, but they did invariably result in one or two per cent greater support from South Asian parents for the maintenance of language.

More conclusive evidence was produced by the 1978 Survey on the aspirations which parents had for their children. 56 per cent of East Africans wanted their daughters to remain at school, 5 per cent preferred them to stop at home and 5 per cent wished them to marry. These responses can be compared with those from the South Asian parents: amongst them, 44 per cent thought further education would be desirable, 14 per cent preferred marriage and 14 per cent thought that their daughters should remain at home. Clearly then, in comparison with the East Africans, the South Asians exhibit a greater tendency towards traditional beliefs, particularly with regard to the role of women, whilst the East Africans might be thought to be pursuing more modern strategies. The same conclusions can also be drawn from the desires of parents regarding the form of their children's dress. In the 1978 Survey, parents were asked to state how frequently their children wore traditional (i.e. non-western) dress. The results showed that around 19 per cent of East African children always wore traditional dress whilst nearly 27 per cent of South Asians did so.

Whilst many of the data presented above are indicative, they are by no means conclusive. Other aspects of research in the sphere of education are less equivocal, but they relate almost solely to the South Asian population. They do, however, show that encapsulation, and therefore restricted interaction between ethnic groups, is also characteristic of the life-style of children as well as of adults. In Blackburn, the Borough pursues a more enlightened policy than is the norm since it allows parental choice of primary schools rather than operating a rigid catchment-area system. Parents can thus choose whether their children attend a school where the majority of pupils are also Asian, or they can opt for attendance at a suburban school where there are few other Asians. This policy has generated considerable hostility among suburban whites and it was this reaction which forced the Borough to cease compulsory dispersal by 'bussing' and move towards the current voluntary system. Although one has to admit that the constraints of distance and time will deter many parents wishing to send their children to suburban schools the avail-ability of free 'bus-passes' and the fact that many parents do successfully send their children by public transport show that the constraints are by no means insurmountable. In the light of the knowledge that such alternatives exist it is indicative to discover that segregation between white and Asian children in Blackburn's primary schools was as great as 62 per cent in 1980 (calculated over 53 schools with 12,291 white children and 1642 Asian children). Further-more, although Asians form only 11.8 per cent of the primary and infant school population of the town, they do form a majority in at least three schools and in one of these, they constitute 65 per cent of the pupils. Stated in other terms, $_aP^*_a$ (or the probability that an average Asian child is likely to interact with another Asian at school) is 0.37 when calculated for the same data set as the ID. This compares with the expected probability of 0.11 if the Asian population was unsegregated. Furthermore, the combination of 1977 Asian Census data and the 1980 educational data allows an analysis of patterns of intra-Asian separation. Over those 29 primary and infant schools which had Asian children on their rolls in 1980s, the ID between Gujurati-speakers and Punjabi-speakers, proved to be 40. It would appear that, at the very least, Asian parents are content to let residential separation follow through into separation at school, whilst they may in fact be consciously choosing to perpetuate this.

Whilst it has been mentioned above that Asian children appear to be encapsulated on a school-by-school basis, it is also important to discover whether the same is true within schools. In other words, do those parents who consciously choose to send their offspring to suburban schools in order to encourage mixing with whites actually achieve this? With this in mind, sociometric testing was undertaken in a large primary school in Blackburn. Whilst no claims to universality can be made on the basis of such a small sample, the techniques which were used are well-tried (see Northway, 1952 for a discussion of methodology). The three-choice three-criteria test was used, whereby children were asked to name the individuals with whom they most liked to play at school, with whom they preferred to sit in the classroom, and with whom they most liked to work. In each case respondents were asked to list three individuals in rank order. Tests were given to three classes of nine-and ten-year-olds. These yielded 69 completed responses, and 621 bits of information concerning patterns of preference within and between white and Asian peer groups. The percentage which Asians formed of the register in these three classes varied between 50 per cent and 77 per cent.

The results of this testing revealed the extent of South Asian encapsulation. As figures 9.5 and 9.6 demonstrate, even where schools or classes are ethnically mixed, both interaction and preference are compartmentalized and encapsulated. In each of the diagrams, white children are found to the left of the vertical divide and Asian children to the right. Linkages between individuals are shown where a child selected another child as one of his nine possible choices on any of the three criteria. Arrowheads reveal the presence or absence of reciprocity, and the relative centrality of an individual indicates his or her overall popularity within the peer group. Although cross-ethnic preferences do exist, the over-

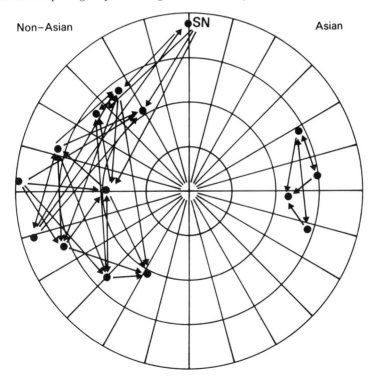

FIG 9.5: Three-choice, three-criteria target sociogram; Class 14 girls

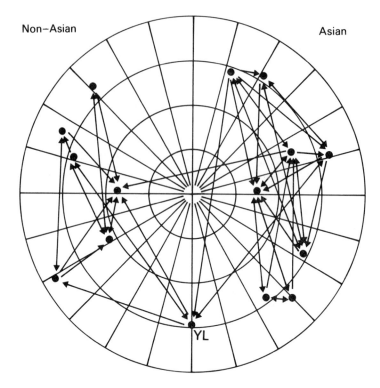

Non–Asian

Asian

YL

FIG 9.6: Three-choice, three-criteria target sociogram; Class 14 boys

riding pattern is one of separation, with a dominance of close-knit, reciprocal preferences within the two groups. Analysed in a different form, of the nine choices given to each Asian child (three choices for each of three criteria), 87.4 per cent were for other Asian children. In more detail, 82 per cent of Asian children selected only playmates of their ethnic group, 68 per cent selected only in-group peers with whom to sit, and 58 per cent selected only fellow ethnics as work-mates. These percentages are considerably higher than comparable results reported elsewhere. Davey and Norburn (1980) undertook a similar analysis of a sample of 238 Asian, West Indian, and white children drawn from schools with large ethnic populations in Yorkshire and the South. They found that expressed friendship patterns were least often solely in-group in character amongst West Indians (21.5 per cent) and most often in-group amongst Asians (44.6 per cent). Davey and Norburn's findings on patterns of friendship amongst white children are also divergent from those discovered in Blackburn. They state that almost 64 per cent of white children had friendships exclusively restricted to their own ethnic group. The comparable figure for Blackburn was 90.3 per cent.

Unfortunately the Blackburn sample contained only one child of East African Asian origin, and one child of a mixed white–Asian marriage. Despite this, these two children, YL and SN, displayed considerable variety in their choice of partners, selecting members of both ethnic groups without favour. This feature is clearly apparent in the three-choice, three-criteria graphs. However, figures 9.5 and 9.6 are sociometric portrayals of those links of preference which exist only *within* school classes. Scrutiny of the network of *all* friendship preferences (including those directed to individuals in other school classes) reveals that ethnicity and group consciousness appear to unite South Asians from different school classes to form a larger community. In contrast, white

THREE-CHOICE LINKAGE PATTERNS ON CRITERION OF
IN-SCHOOL PLAY CLASS 14 AND CLASS 12 BOYS, ASIAN

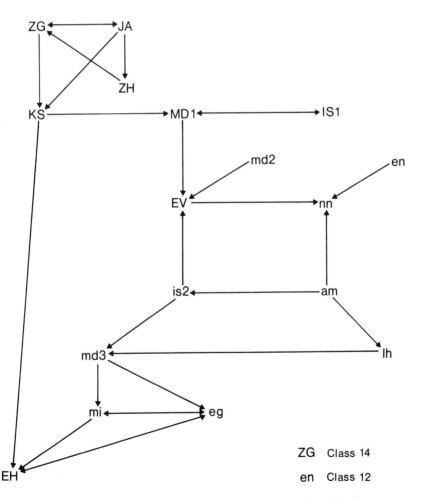

FIG 9.7: Three-choice linkage patterns on criterion of in-school play; Asian boys

children select primary-group contacts largely from within their own school classes; this consequently produces several mutually exclusive social networks rather than a larger unified body (see figures 9.7 and 9.8). In the case of Asians, over 18 per cent of first-choice primary contacts were amongst Asians in another school class. South Asian children thus became encapsulated not only within the classroom but also within the school as a whole. One could argue that this results from a poorer level of spoken English which prevents contact with indigenous children. This, however, could be contradicted by the ability of certain children to create primary-group contacts across ethnic lines where they wished to do so.

Perceptions

The aim of this chapter so far has been to demonstrate the way in which the contrasting purposes and permanence of East African and South Asian migration to Britain become translated into differing behaviour once the migrants have arrived in this country. This final section aims to explore whether the attitudes which the migrants possess upon arrival have influenced the experiences which they have encountered in Britain, and how both these factors have shaped their perceptions of British society and its inhabitants. It is likely that these perceptions will, in turn, challenge or reinforce behavioural patterns.

It has been suggested that East Africans are more willing to compete with white members of society for access to resources and that this attitude stems from higher aspirations and a belief in permanent settlement. The 1978 Survey specifically collected information relating to the use of services and satisfaction with these, and these data tend to confirm the argument. When asked for an opinion on the effectiveness of the local Community Relations office, 72 per cent of East Africans were able to proffer one on the basis of personal experience. The same could be said of only 62 per cent of South Asians. When asked about the use of medical facilities (the family planning clinic, dentist, GP, or ante-natal classes) 12.5 per cent fewer East African women admitted to never having used any of these facilities or to only have used one. Conversely, 25 per cent more East Africans than South Asians had used two, three, or four of these services regularly. Finally, when asked about the effectiveness of the police and their attitude to ethnic minorities, only 5 per cent of East Africans as opposed to 14 per cent of South Asians described them as very helpful. On the other hand, 22.3 per cent of East Africans thought the police to be unconcerned or actively hostile, against 17.5 per cent of South Asians. However, willingness to compete for resources entails contact not only with institutions but also with individual members of the core culture. One might therefore reasonably expect East Africans to have had greater primary-group contact with white Britons either in the workplace or the area of residence. Again, the 1978 Survey shows this to be the case. When asked how many British friends they had 16.7 per cent of East Africans replied 'many' and 16.7 per cent replied 'none'. Amongst the South Asians, 7.5 per cent thought they had many British friends and 29 per cent thought they had none.

THREE - CHOICE LINKAGE PATTERNS ON CRITERION OF IN-SCHOOL PLAY CLASS 14 BOYS, WHITES.

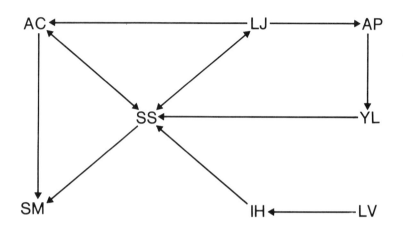

THREE-CHOICE LINKAGE PATTERNS ON CRITERION OF IN-SCHOOL PLAY CLASS 12 BOYS, WHITES.

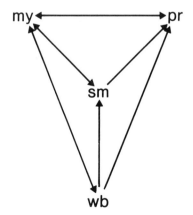

FIG 9.8: Three-choice linkage patterns on criterion of in-school play; white boys

Whilst Gordon's (1964) influential model of assimilation sees primary-group contact as the first step towards harmonious assimilation, it can also be argued that greater contact simply alerts minority members to the strength of feeling against them and to the existence of those structural forces which bar their social mobility, and which were discussed in chapter 4. Such an interpretation seems to be the more valid one for those East Africans interviewed in Blackburn. Almost 18 per cent more South Asians than East Africans described people in their street as 'very friendly'. More than twice as many South Asians as East Africans described British people, in general, as being 'very friendly' whilst 32 per cent more East Africans used the description of 'withdrawn' or 'hostile'. Fourteen per cent more South Asians named the British as their most preferred ethnic group and 11.8 per cent more East Africans thought that the English were their least preferred. However, perhaps more important than any of these issues is the question of whether East Africans believe more fervently that racial and ethnic discrimination occurs in British society and whether they have any experience to support their beliefs. Again, the 1978 Survey data seem fairly clear on this. When asked if they agreed with the statement that British people treated them differently because of their colour, 19 per cent of East Africans agreed strongly and a further 39.2 per cent agreed. Less than 25 per cent disagreed with the statement and no-one disagreed strongly. Amongst South Asians, though, 14.6 per cent agreed strongly and 34.9 per cent agreed whilst 32 per cent disagreed and 1.2 per cent disagreed strongly. The CRC's (1977) study in Bradford also demonstrated this belief amongst South Asians that discrimination is not widespread. Finally, when asked if they themselves had suffered discriminatory actions, and if so, how often, a full 30 per cent more East Africans in Blackburn replied that they had experienced discrimination either 'very frequently' or 'frequently'. Again this discrepancy was picked up by the CRC (1977) report.

Summary

This chapter has considered whether the contrasting pre-migration experience of East Africans and South Asians which was discussed in chapter 7, and the differing motives for migration (discussed in the same chapter) have produced concomitant differences in the behaviour of the two groups and their perceptions of British society. It seems that they have. The South Asians to which the Blackburn data related have restricted aspirations whilst in the United Kingdom, centred upon frugality and the myth of an early return to the sending society. When this is allied to structural constraints it has ensured that the group remains encapsulated in its neighbourhood of residence, in its place of work, and in its place of learning. Despite objectively inferior conditions, limited contact with the standards of the core culture and the persistence of pre-migration values produced a situation where South Asians are less conscious of their relative deprivation. Their desire to maintain boundaries by introspection and the retention of close-knit social networks (designed to minimize cross-ethnic primary-group relations) shelters South Asians from the full force of

hostility and inequality which exist in the wider society. South Asians consequently hold a more idealistic view about the attitudes of British society
towards them (although this may well change under the pressure of economic
recession and violent attacks), and moreover, they remain unaware of many of
the exclusionary constraints which would bear upon them if they decided to
venture more widely into British society.

In contrast, those East African Asians studied here (and it is important to
recall that our sample appears more conservative than most) have less
restricted aspirations, a built-in drive to social and economic mobility, and a
more limited desire to retain community association. They also appear both
more willing and better able to engage in exclusionary closure against local
whites. This willingness and desire to compete, and the removal of cultural
defences which this necessitates inevitably exposes them to the full extent of the
constraints which bear upon them, and consequently alerts them to the overt
hostility of the core culture. Unprotected by an encapsulated and therefore
insulated society and unable to gain full acceptance to the core culture because
of their racial origin and ethnic background, East African Asians seem set
on a route which may well end in increased inter-ethnic conflict and group
marginality.

CHAPTER 10

A Continuum of Ethnic Association

Previous chapters have concluded that white society undoubtedly engages in persistent and significant exclusionary closure against the Asian population but that the impact of this depends upon the strategies adopted by Asians and the purpose for which they sought residence in Britain. Other chapters indicated that there was no such thing as 'the Asian' population but rather two main minorities of Asian descent, the South Asians and the East African Asians. However, whilst the category 'Asian' is in reality only a convenient mental construct used by the white population, the notion that the terms 'South Asian' or 'East African' have any greater empirical reality is also false. Chapter 9 has already shown that 'the South Asian' population is actually an amalgam of different ethno-linguistic groups, each of which is attempting to maintain its social distance from the others, not only at home, but also at work and at school. A visit to the major sending areas confirms this belief. The Muslim taxi driver in Gujarat who bemoans his inability to get a good job or house because of Hindu discrimination, the Pakistani Muslim who takes unconcealed delight in describing how Hindus ascribe medicinal properties to cow dung, the perpetual but conflicting claims about Indian and Pakistani military prowess, and the persistent criticism of Muslim polygamy by Hindus all lead the field researcher to the conclusion that 'South Asian', 'East African Asian', and 'Indian' may well be convenient labels but that they are also irrelevant. Physical conditions in the different parts of the the sub-continent further underscore this belief. Life appears very different when viewed from the balcony of an elegant *pakka* house in the fertile Jullundur district than it does when seen from a *kutcha* house built in a poverty-stricken village in the arid and barren wastes south of Rawalpindi. The contrast becomes even sharper if comparisons are made with urban Nairobi or Harare.

Despite the acknowledged existence of these internal differences within Britain's Asian population, little systematic research on this topic has yet been published. Clearly, one of the reasons for this is that the decennial census does not collect data on such fine-scale groupings, and researchers are therefore forced to collect their own. Given the need for large samples and the requirement for data with great depth, the paucity of published work is understandable. However, the work which has appeared has been both interesting and consistent in its conclusions. Brooks and Singh (1979a), for example, analysed an unspecified sample of Asians drawn from manufacturing industry in the West Midlands. The authors were particularly interested in the extent to which groups had remained 'traditional', and whether they displayed different degrees of commitment to permanent settlement in the UK. They concluded that a 'four-stage typology' existed with traditional groups at one end and non-traditional at the other. The East African Asians were the least traditional, followed by the Gujurati Hindus and the Punjabis (Sikhs). The Pakistanis were the most traditional. This typology received support from a study of Asian retailers undertaken in Coventry (Robinson and Flintoff, 1982). Whilst the sample used in this analysis was admitted to be small, the results proved to be internally consistent. The authors found that the Gujurati Hindus (most of whom were East Africans) had achieved considerable economic success within the confines of the ethnic market; that Pakistani Muslims had maintained a high degree of social encapsulation but had been the least economically success-ful; and that the Punjabi Sikhs had been willing to cater for a non-ethnic clientele in order to achieve economic targets. Baker (1982) produced similar conclusions from her study of Asian businesses in London.

This chapter pursues the theme of internal differentiation but it does so in greater depth than previous published studies. In particular it seeks to present a continuum of ethnic association along which the various ethno-linguistic groups can be arranged. This continuum stretches from almost total encapsu-lation at one extreme to desired, but blocked, assimilation at the other. In consequence it sheds further light on whether groups are, in reality, adopting stances predicted for them in the model outlined in chapter 6. However, before such analysis can be undertaken, it is important to state various caveats. Firstly, any continuum of this type can never be divorced from the circumstances which surround it. These are assumed to be characterized by racial hostility. Secondly, the conclusions of this analysis relate only to Blackburn and those ethno-linguistic groups which reside there. The conclusions may have validity outside this geographical context but this is not proven here. Thirdly, as table 10.1 indicates, some of the Blackburn groups are small in number, and conclusions about these groups can only be presented as being suggestive. Having said all this, table 10.1 provides a detailed breakdown of the make-up of Blackburn's Asian minority. It is immediately apparent that this is dominated by individuals from rural as opposed to 'urban' backgrounds. Furthermore, the minority is divided into two major components; Punjabi speakers (29 per cent), and Gujurati speakers (59 per cent). This division is cross-cut by addi-tional affiliations, most notably those relating to religious beliefs (largely Muslim but with a significant Hindu component) and experience of life in East Africa. In total, sixteen ethno-linguistic groups were represented, but of these, four were too small for meaningful analysis. Other groups, such as the Sikhs, have been included although their numbers allow only tentative conclusions to be drawn.

Ethnic Association

The topic of ethnic association, and the process by which this association may be gradually eroded (assimilation), has been central to the sociological and demographic literature for some considerable time. Naturally, different disci-plines and different workers take contrasting approaches, and as a result the literature is both extensive and broad. Price (1969) provides one of the best summaries of this diversity. He groups those works which look at the process of assimilation into three: firstly, those which simply describe the degree of assimilation of a given group (see Gans, 1962 for example); secondly, those which attempt to classify, by the creation of typologies (e.g. Petersen, 1970), by the comparison of migrant groups with their respective host populations (e.g. Lancaster Jones, 1967), or by the construction of temporal sequences of assimilation against which the progress of any one group can be compared (e.g. Wirth, 1928); and thirdly, those which generate formal theories or models (e.g. Bogardus, 1930).

One of the most influential of these many works on assimilation is the model proposed by Gordon in 1964 and subsequently modified by him in 1975. Gordon argued that the process of assimilation begins either with structural assimilation or acculturation. The former can be defined as minority entry into large-scale primary-group contact whilst the latter is the adoption by minority-group members of the cultural traits of the host population. Gordon implied that, of these two changes, structural assimilation was the more important, since this automatically produced acculturation. The same could not always be said in reverse. He also made the important distinction between two different levels of acculturation: a group would initially adopt the extrinsic cultural traits of the core culture (such as dress, accent, or manner) and would only later take up the intrinsic cultural traits such as religion, and ethical values. Once group members had become both structurally integrated and acculturated, the process of assimilation would progress to the next phase, namely large-scale outmarriage. Such exogamy would then encourage the minority to identify with the core society and assimilation would therefore have proceeded a further stage. The final phase is dualistic in the sense that it encompasses not only changes in the attitudes held by majority members (i.e. a decline in prejudice) but also changes in societal structures so as to eradicate power conflicts. When a minority has reached this point it can be said to have a high degree of 'total assimilation' (Gordon, 1975).

Towards a Reformulation

Although Gordon's model is perhaps the most robust of those available, it has not gone uncriticized. Yinger (1981), for example, has noted that the model only concerns itself with group assimilation and therefore assumes that individual assimilation follows a similar course. In practice, this is not often the case. Yinger also argues that the term 'structural assimilation' is too broad since it covers so many different types and forms of interaction. Price (1969) has criticized the inevitability with which assimilation is assumed to take place once structural assimilation has occurred. Whilst such thinking was very much in line with the 'conventional wisdom' (see Kantrowitz, 1981) of the time, it takes no account of the underlying barriers which might halt a group at any point in the assimilation process. Nor does it allow for regression. Further, the model cannot accommodate the fact that different minority groups will have differing aspirations depending upon the motives for migration. Whilst

Gordon's structure may well be ideal for the circumstances of American immigration in the nineteenth and twentieth centuries it is less than perfect for the temporary economic movements and refugee crises which have characterized European migration since the Second World War. It is also possible to criticize Gordon at a more theoretical level. Whereas in 1964 Gordon had implied an ordered progression through sequential phases of assimilation, by 1975 he had retreated from this stance and instead offered only an unordered check-list of assimilation criteria. To suggest that assimilation takes place in this way under-values the causal links which exist between stages and denies much of the value of the process of modelling. Finally Gordon omits any mention of spatial pattern either as a cause of social attitudes or as an outcome of these. In the light

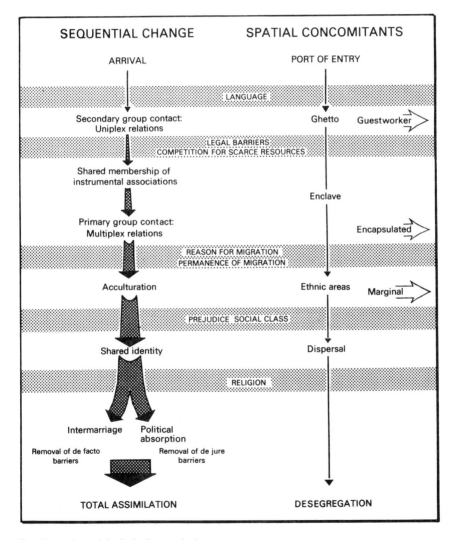

FIG 10.1: A model of ethnic association

of these various criticisms, it would seem appropriate to formulate a new model, but one which is still based very much on Gordon's original ideas.

Such a model is contained in figure 10.1. It reverts to Gordon's earlier notion that assimilation might take place in a sequence of identifiable stages. It does not however assume that assimilation is inevitable, and it places a good deal of emphasis upon the barriers which might prevent the attainment of this goal and which might instead produce a strengthening of group identity. The first barrier which a group must surmount relates to the issue of communication. Without at least a basic ability to communicate with members of the core society, a minority will not be able to gain employment and the secondary-group contacts which are associated with this. They will remain sealed in their port of entry. However, once individuals are able to communicate at a basic level, they may pass on to the second stage where they gain employment in the core society and are able to make neighbourhood contacts. In both of these cases, though, relationships are single-stranded or uniplex in nature and need not imply anything other than functional contact. The next barriers which a group faces are legal restrictions on social interaction (e.g. under apartheid) and the fear held by members of the core society that the minority is seeking to usurp scarce resources. If these barriers prove impenetrable, a group is consigned to become merely guest-workers who fulfil an economic need and who are constrained to live in ghettos (or even company dormitories). Conversely, if a minority *can* persuade members of the core that it is not a competitive threat, and if legal barriers to social interaction are either dissolved or do not exist, it will gain access to instrumental associations (trade unions and neighbourhood bodies) and the primary-group contacts which these may stimulate. These contacts will be multiplex in character and carry with them some emotional content. However, the development of such contacts raises into prominence the aims and objectives behind the migration of the minority. If a group has sought migration purely as an economic expedience and does not foresee permanent settlement in the core society, it is unlikely to allow primary-group contacts to develop to the extent that they threaten its culture and its prospects for return migration. Encapsulation will ensure that the group retains its own culture and seeks no further assimilation. If however the aims and objectives of the minority are compatible with acculturation then the group will pass on to this stage. For those who are encapsulated, the enclave will develop as a long-term spatial symbol of their cultural independence. For those willing to become acculturated, the enclave will form only a temporary phase.

Having adopted the culture of the core society, it would seem natural for a minority also to take on the identity of that society and for members of the majority to accept them progressively as equals. However, whilst this may be a 'natural' course of events, it is not always the one which occurs. Prejudice may alienate the minority, whilst pre-existing stereotypes may prevent the majority from regarding them as equals. In addition, a convergence of identity will also be retarded if the minority has a skewed and inferior social class profile. In these cases, the minority may well be deflected from an assimilationist course and instead become a marginal group with frustrated aspirations. The spatial concomitant of the acculturation phase is the ethnic area: this is usually a fairly long-lived phenomenon and often takes the form of a microcosm of the broader society with many of its institutions, facilities, and services. This 'institutional completeness' (Breton, 1964) is the characteristic which differentiates the ethnic area from the ghetto.

Finally if a group does succeed in taking on the identity of the core culture and being accepted as equals, the way has been paved for its members to engage in exogamy and for the group itself to be legally recognized as equal. The former marks the end of *de facto* barriers whilst the latter signals the removal of the *de jure* barriers. The minority could also become spatially integrated through dispersal from ethnic areas and this process would provide positive feedback for even greater outmarriage and for more complete political integration. The minority would have achieved total assimilation and be indistinguishable from the remainder of society (unless biological differences remained).

Relative Ethnic Association within Blackburn's Asian Population

Having described a proposed assimilation chronology, it remains to employ this tool as a way of highlighting those ethno-linguistic groups within the Asian population of Blackburn which are set on an assimilationist course, from those which are maintaining their encapsulation. It will then be possible to compare the Blackburn results with those from the other published typologies mentioned at the start of the chapter. The purpose of this section is to begin the analysis by looking only at relative differences within the minority. No reference will be made to absolute levels of assimilation until the next section.

The 1977 Asian Census and 1978 Sample Survey both provide data which are relevant to the task of operationalizing the model of ethnic association outlined above. Surrogate variables are available to indicate the potential which a group has for assimilation. Other data quantify the degree to which groups have surmounted these obstacles and passed from one assimilation phase to the next.

Secondary-group contact

Whilst structural integration has often been seen as the key to the process of assimilation, it is important to distinguish between the early stages of integration which involve entry into secondary-group contact through both the workplace and the neighbourhood of residence, and the later stages which consist of penetration of primary-groups through membership of associations or clubs. It is for this reason that these two different phases of structural integration were separated so as to form independent stages in the model.

One of the most frequently used indicators of secondary integration is a comparison between the industrial and occupational structure of minority groups and the majority population. Examples of the use of this approach include Duncan and Lieberson (1959), Lieberson (1961), and Taeuber and

Taeuber (1964). The thinking which lies behind it is that new migrants initially take jobs of a low economic status because of their lack of knowledge, shortage of skills, and need for money. Only as they become more acculturated and established do they seek the occupational mobility which breaks down the early specialization of the group into those niches abandoned by the majority population. In short, the level of economic diversification of a group's employment structure is seen as a direct measure of how far that group has progressed along the assimilation continuum. Clearly, though, this ignores the fact that groups may sometimes only locate a vacant economic niche which they can exploit after a lengthy period of residence in a new society. In this case, economic specialization may actually increase through time.

Other authors argue that economic diversification has a greater significance than acting as a simple assimilation indicator. Hechter (1975; 1978), for example, considers that the existence of a division of labour along ethnic or cultural lines may allow a group to maintain its identity despite the increasing uniformity and conformity enforced by economic modernization. This is particularly true where the division of labour is consciously maintained by the dominant group through the manipulation of cultural symbols, the use of stereotyping and the media, and overt exclusion. Yancey *et al.* (1976) also support the view that economic specialization has a more active role to play in maintaining ethnic solidarity. They argue that group identity receives support from five sources in the occupational sphere. First, similarity of occupational status is often linked with common economic status and a commonality of life-style. Second, shared occupational status may well be linked to class consciousness and therefore to group solidarity. Third, shared occupations provide individuals with common social and economic interests. Fourth, industrial concentration provides the basis for interpersonal relationships and therefore for primary-group contact. And finally, the tendency of people to travel to work only over relatively short distances means that workplace segregation often follows through into residential segregation.

Occupational segregation is, however, not the only measure of a group's structural integration through secondary-group contact. As was argued in an earlier chapter, residential segregation is also an important indicator. The assumption here is that patterns of residence reflect the social distance which exists between groups, and therefore the likelihood that group members may become involved in shared associations and friendship circles. It is, however, important to note that an absence of spatial distance may not always produce greater harmony and interaction but instead may result in discord and conflict. Rex and Moore's (1967) findings in Sparkbrook are a case in point.

The 1977 Survey yielded data on the industrial structure of each of the twelve ethno-linguistic groups. Again, the Gibbs–Martin Index of Diversification was used to create a single summary statistic of relative concentration or diversification. However, because of the acknowledged sensitivity of this index to variations in the number of industrial categories (see Clemente and Sturgis, 1971), the calculated index was modified to represent a percentage of the maximum diversification possible given the number of categories in which a group was represented. A figure of 100 per cent therefore indicates maximum diversification whilst 0 per cent represents complete concentration.

The residential patterns of the twelve groups were also summarized, using the index of dissimilarity for households at the ward level. Initially, IDs were calculated for each group against the white population but the same basic information can also be used to study patterns of fission *within* the Asian minority. It is best to consider these first, and they are therefore presented in table 10.2. What this reveals is that levels of residential segregation vary considerably between the 20.4 minimum (Urban East African Gujurati Muslims–Urban Indian Gujurati Muslims) and the maximum of 85 (Urban Indian Gujurati Muslims–Indian Gujurati Hindus). In order to highlight the overall pattern of segregation, a target sociogram was produced from the matrix of IDs such that the three lowest IDs scored by each group were taken as indicating those three other sub-groups with which residence was preferred. It should be noted that the sociogram portrays only those patterns of preference which were reciprocated. In other words a link labelled as being 'first order' represents a situation where sub-group A is less segregated from sub-group B than from any of the other ten groups, and vice versa. Although the sociogram (figure 10.2) ignores the absolute level of segregation, it does clearly demonstrate that, at this level, there are five major elements within the Blackburn

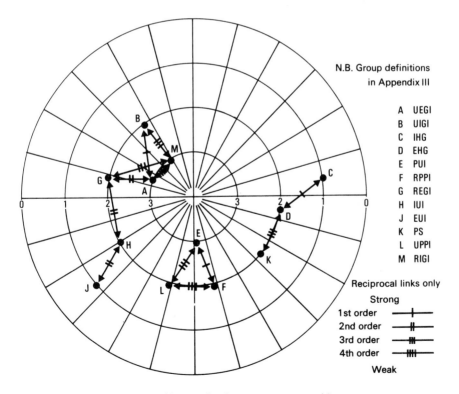

N.B. Group definitions
in Appendix III

A UEGI
B UIGI
C IHG
D EHG
E PUI
F RPPI
G REGI
H IUI
J EUI
K PS
L UPPI
M RIGI

Reciprocal links only
Strong
1st order
2nd order
3rd order
4th order
Weak

FIG 10.2: Target sociogram of interaction between groups; residence

population. The two most important of these are the Gujurati Muslims (A,B,G,M) and the Pakistani Muslims (E,F,L). Within the former, first order links associate the two urban groups (A,B), second order links associate the two East African groups (A,G), and third order links associate the two Indian groups (B,M), and the rural Indians with both East African groups (M,A, and M,G). Within the Pakistani Muslim cluster, first order links associate the rural Pakistani-speakers with Urdu-speakers (E,F) and third order links associate the two Pakistani Punjabi-speaking groups (F,L), and the Pakistani Urdu-speakers with the urban Punjabi-speakers (E,L). Outside these two main cluster, satellite clusters also exist. The Indian and East African Urdu-speakers (H,J) are associated, and appear as a satellite of the Gujurati Muslim cluster. The two Hindu groups (C,D) are mutually associated by a first order link and have a weaker tie to the Sikhs (K). The Punjabi Sikhs themselves are somewhat isolated, but are also associated with the Urdu-speaking Punjabis (E). If one calculates the average ID between the groups in each cluster, the Gujurati Muslim value is 26.4 between the five groups, the Pakistani Muslim value is 30.3 (between three groups), the Gujurati Hindu value is 33.8 (two groups) and the Urdu Muslim value is 42.8 (two groups). All of these intra-cluster IDs are below the average for the whole matrix of 66 IDs, which proved to be 50.9. This indicates that the levels of social distancing within the clusters are considerably less than those which exist between them. The entire analysis underlines that it really is essential to break the Asian minority down into such small fractions rather than attempting to generalize about the whole population. Such generalizations would not only be meaningless but also misleading.

To return to the *external* relations of the various Asian groups though, table 10.3 summarizes the findings on secondary-group contact (see columns 2 and 3). The groups which enjoy the greatest degrees of secondary integration are the Punjabi Sikhs, the East African Urdu-speakers, the three East African Gujurati groups, and the Pakistani Urdu-speakers. The groups which are the least integrated are the Indian Gujuratis and the Pakistani Punjabis. Given the conclusions about East Africans and South Asians presented in previous chapters, these results do not seem surprising, although it is worth pointing out that Urdu is a language of the urban middle classes, and that these groups therefore might also be expected to evidence greater modernity.

Associations and primary-group contact

Membership of associations and clubs represents an important stage in the assimilation chronology. It brings individuals from different ethnic groups together, it gives them a shared purpose, and it allows people to form their own opinions about each other as individuals rather than relying upon second-hand stereotypes (see Miller and Brewer, 1984). Contacts gained through clubs and cliques also provide the basis for primary-group contacts to develop with their greater emotional content and multiplex character. Lastly, shared membership of associations gives minority-group members the opportunity to observe the cultural characteristics of the core culture in an uncompetitive atmosphere which is conducive to acculturation.

The 1978 Sample Survey collected data on levels of trade union membership amongst Asians in Blackburn. This is by no means an ideal measure of the extent of shared membership of associations since it is an acknowledged fact that trade union membership is higher in larger plants and branch factories of national companies than it is in small plants owned by local entrepreneurs. Since it has already been demonstrated that South Asians are found to a greater extent in the former of these categories, it would also be expected that their levels of union membership would be higher, regardless of whether they were more assimilated. Membership of local branches of political parties might have been a better indicator but information was not collected on this topic in 1978. Alternatively the analysis could have concentrated upon the relative absence of membership of Asian associations. Nevertheless, the data on union membership are presented in the summary table, table 10.3. This shows that four groups enjoyed relatively high levels of secondary contact: the Sikhs; the Indian and East African Urdu-speaking Muslims; and the East African Gujurati Hindus. Low levels of secondary contact were a feature of all four Gujurati Muslim groups and Pakistani Urdu-speakers. Whilst many of these results might have been predicted, the absence of secondary-group contact amongst the Rural and Urban East African Gujurati Muslims is somewhat unexpected. This issue is returned to later.

Whilst membership of an association *can* provide opportunities for primary-group contact, it remains to be seen to what extent these opportunities are realized in the form of multiplex minority–majority relationships. A question in the 1978 Survey asked respondents how many British friends they felt they had. Here, the analysis concentrates only upon those individuals who thought they had no British friends. The range of values proved to be quite large, from the Punjabi Sikhs at one extreme (none of whom lacked British friends) to the Urban Indian Gujuratis at the other. It is noteworthy that the East African groups generally exhibited a greater tendency towards having British friends. This conclusion parallels that of the Community Relations Commission (1976).

Acculturation

Acculturation, along with structural integration, is usually regarded as one of the first phases of assimilation. However, it is a very broad term which can mean the deletion of a characteristic element of an individual's diet at one extreme, to the wholesale internalization of a new value-system by the entire group at another. The concomitant of this is that general definitions of acculturation are rarely very valuable. However, despite this diversity, most definitions share some common ground in that they stress the following points: smaller or weaker groups become acculturated more rapidly than large ones because they do not have the mass to allow institutional completeness; acculturation is more rapid in societies where prejudice and discrimination are absent; and different elements of the core culture are adopted by a minority at different points in time. This last finding and the earlier comments about the breadth of the concept, indicate the need to disaggregate the notion of acculturation. Gordon (1964) distinguished between intrinsic cultural traits (such as religion, literature, and shared past) which would be more difficult to adopt and extrinsic cultural traits (e.g. accent, dress, or manner) which could be emulated readily. It is also possible to separate substitutive cultural traits which replace minority ways, from additive cultural traits which simply augment the old ways. Recently, though, the emphasis which analysts have placed upon the whole of acculturation has begun to wane for two major reasons. First, it is now realized that becoming acculturated with regard to extrinsic cultural traits may not simply represent a temporary phase in the movement towards total assimilation, but rather a more permanent and long-lasting state beyond which a group may not move. The model presented in this chapter takes account of this. Second, cultural variables may not have the significance which early analysts gave them. They can be manipulated in order to strengthen instrumental movements without this representing any change in the views of group members (Bell, 1975), or alternatively, cultural variables may simply express a 'symbolic ethnicity' (Gans, 1979). Gans argues that the use of ethnic symbols may reflect a need for identity rather than a desire to belong to a functioning and coherent cultural group. The adoption of cultural signs is thus a leisure pursuit rather than a deep commitment.

The 1977 and 1978 data sets contain information on the adoption of both intrinsic and extrinsic cultural traits. The latter were represented by data on the frequency with which adult women and children under the age of eleven years wore 'traditional dress'; in particular, the percentage who wore traditional dress at all times. These data produced the rank ordering shown in summary table 10.3. The East African Urdu-speakers were again shown to be a relatively acculturated group, as were the Sikhs, the East African Gujurati Hindus, and the Rural East African Gujurati Muslims. Low levels of acculturation were recorded by the Rural Punjabis, the Urban Indian Gujurati Muslims, and the Urban Punjabis (cf. Community Relations Council, 1976). The intrinsic cultural traits were represented by the issue of religious faith. None of the 8862 people interviewed in the 1977 Asian Census had given up their own religion to become Christian, although several individuals from Goa were already Catholics prior to their arrival in Britain. A rank order of relative intrinsic acculturation could not be constructed.

Identification

The extent to which the ethno-linguistic groups thought of themselves as 'British' provided a good measure of their identificational assimilation. The 1978 Sample yielded data on two separate aspects of this: firstly, the groups' opinions about the friendliness or hostility of British people; and secondly, the results of a crude social-distance scale in which respondents were asked to rank various ethnic groups in order of preference. Other measures which could have been used include the level of political participation of the different groups and their attitudes towards issues of national concern. Analysis of the two variables which were used, however, revealed the relative identificational assimilation of the Sikhs, the Gujurati Hindus and, to a lesser extent, both East African Gujurati Muslim groups. Conversely, all three Urdu speaking groups were ranked in low positions, as were the Urban Pakistani Muslims. This was not all that the data revealed, however. Rank order correlation demonstrates that there is an inverse relationship between the extent to which a group regards itself as close to British people and whether that group views British people as friendly (rs = 0.51). Whilst this relationship is by no means perfect, it suggests that as Asians come to feel closer to British society they are also made increasingly aware of its hostility towards them as coloured people. This is in full accord with the conclusions of previous chapters where it was noted that encapsulated South Asians had a more benign view of British society than their competitive East African counterparts.

Amalgamation

The level of group outmarriage is commonly regarded as one of the ultimate measures of minority assimilation. Marriage represents perfect positive interaction and therefore implies the absence of social distance. In practice, it is now known that intermarriage need not always result in permanent assimilation and that, in certain cases, intermarriage can take place without any other form of assimilation. The children of a mixed marriage do not necessarily become attached to the majority culture: they may form an intermediate group with its own identity or they may simply be absorbed back into the minority group as full members. Similarly, where no alternatives exist to outmarriage, this may take place with little prior acculturation, integration, or identificational assimilation. Thus whilst levels of outmarriage are usually a good indicator of assimilation, this is not always the case.

Data from the 1977 Asian Census allow an analysis of levels of outmarriage to members of the majority culture, and hence the level of biological amalgamation. A detailed study of this data source appears elsewhere (Robinson, 1980d), but in summary, it relates to 1185 marital couples. Each of these was checked for similarity or dissimilarity in the ethnic, or ethno-linguistic, affiliation of groom and bride. The results of this analysis are recorded in table 10.3. It should be noted that the complete lack of exogamy recorded by six groups truncates the rank order at this point. Exogamy was most frequent amongst the Pakistani Punjabi Muslims and the Indian Gujurati Muslims, each of which were characterized in the early stages of their migration by all-male lodging houses. Further analysis of these outmarriages suggests that they did take place during this early period between Asian males and British females, and that they therefore reflect a lack of opportunity for Asian in-group marriages rather than significant assimilation. The East African Urdu speakers and East African Gujurati Muslims also recorded exogamy to members of the core culture.

As with patterns of residential dissimilarity, the intermarriage data lend themselves to a parallel analysis of social distancing within the Asian population.

To achieve this, those marriages were isolated where one partner declared either a different birthplace, first-spoken language, or religion to his/her spouse. The degree of out-marriage by each group to other Asian groups was then quantified by means of Savorgnan's index (see Savorgnan, 1950; and Robinson, 1980d), a summary statistic which takes account of exogamy by both grooms and brides. As in other recent uses of this index, out-marriage was expressed not as a proportion but as a percentage. This allows greater comparability with the index of dissimilarity. The complete matrix of values is presented in table 10.4 and it demonstrates that the degree of intra-Asian exogamy varies considerably between the twelve different sub-groups from a minimum of 0 to a maximum of 39.1. Again, the structure of these marital links can perhaps best be illustrated by the target sociogram. This was constructed in a similar manner to that which relates to residential segregation, and is presented in figure 10.3. Clearly, patterns of intra-Asian exogamy do not exactly replicate patterns of residential separation but there is a good deal of overlap between the two. The Gujurati Muslim cluster (groups A,B,G,M,) is still intact, although in the context of marriage it lacks the link with the Indian and East African Urdu-speaking Muslims (groups H,J). This latter pair instead appear to associate with other Urdu speakers and, via them, with Punjabi Pakistanis. In the marriage analysis, the Urban Punjabi Pakistanis seem isolated from other Punjabis, whilst the Sikhs are also entirely endogamous. As in

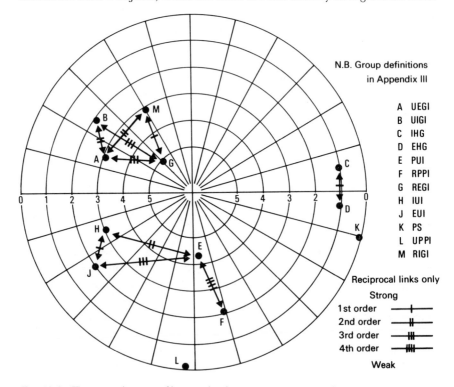

FIG 10.3: Target sociogram of interaction between groups; marriage

patterns of residence, the two Gujurati Hindu groups appear as a self-contained cluster.

Absolute Ethnic Association within Blackburn's Asian Population

The discussion above has been couched in terms of relative degrees of assimilation so that those groups which are more assimilated can be identified from those which are less assimilated. Whilst this approach is valuable in producing a typology of ethno-linguistic groups, it says nothing about the absolute degrees of assimilation of those groups. Although a typology ostensibly arranges groups along the full length of a continuum, in practice it may simply be differentiating a series of groups, each of which is located at a very similar point on the continuum. The detailed process of differentiation may obscure the fact that the groups are objectively very similar. One of the factors which determines the absolute levels of assimilation is the potential which a group has for integration. Some groups can communicate more easily with members of the core society than others. Some groups will be committed to permanent settlement in the society, others will see it only as a temporary economic expedience. Some groups will share the same social-class profile as the core society, whilst others will be markedly over-represented in a narrow band of social classes. Each of these factors determines the extent to which a group might wish to become assimilated and the likelihood that members of the core culture will allow this.

Data are available from both the 1977 and 1978 Surveys to quantify the potential of each ethno-linguistic group for assimilation. The language barrier can be represented by data on the percentage of a group able to speak and write English. The orientation of the various groups can be quantified in three ways: their expressed desire for permanent settlement in the UK; the proportion who still remit money; and the extent to which parents wish to see their mother tongue transmitted to the next generation. The skewness of the groups' social-class profiles can be measured by reference to the percentage which fall into social class 5 (unskilled workers).

The scores of each of the groups on these twelve criteria can be found in rank form in table 10.3 and absolute form in table 10.5. What these data reveal is that the two East African Gujurati Muslim groups, and the East African Urdu speakers all evidenced significant potential for assimilation. The Indian Gujurati Muslims showed less potential and the Pakistanis showed least of all of the large groupings. The Sikhs and Hindus fell between the two extremes. A point of significance is that, in every case where direct comparison was possible, East African groups possessed greater potential than non-East African groups, and urban groups had greater potential than rural groups. In absolute terms, table 10.5 makes it clear that, whilst *every* member of those groups with assimilation potential was not fully equipped for immediate assimilation, the majority were. Conversely, the groups which demonstrated less potential clearly had a considerable way to go before they could be assimilated or before they wish to be assimilated. The Rural Punjabis, for example, had only a bare majority able to speak and write English, a clear majority who were still remitting and who were in favour of maintaining their mother tongue, and more than one in three still wished to return home in the future. Even within a relatively small Asian minority, such as Blackburn's, it therefore seems as if the potential for assimilation varies a good deal.

Secondary-group contact

Reference to table 10.5 reveals the varying absolute values of secondary-group contact experienced by the ethno-linguistic groups. Values of industrial diversification ranged from 55 per cent of the maximum at one extreme to 92 per cent at the other. These compare to the figure for the white population of the North West Economic Region which was 90 per cent in 1977. What is clear, though, is that some ethno-linguistic groups had achieved considerable industrial diversification whilst others had been unsuccessful. The indices of residential dissimilarity reflected a similar trend. Some groups were relatively dispersed (lowest ID 53) whilst others were a good deal more concentrated (highest ID 74). Again this underlines the fact that certain groups (such as the East African Urdu speakers and the East African Gujurati Muslims) had succeeded in gaining significant secondary-group contact whilst others had not.

Associations and primary-group contacts

Trade union membership also varied considerably within the Blackburn sample. The lowest rate was 60 per cent, recorded by the Pakistani Muslims who speak Urdu. The highest was 100 per cent, recorded by four different groups. These figures can be compared with the results of Brown's (1984) national survey. He found that 56 per cent of ethnic respondents and 47 per cent of white respondents were union members. It is important to note that, whilst in Blackburn, the two East African Gujurati Muslim groups might have been expected to have had considerable associational intercourse with the core society, in practice they did not. A possible explanation for this anomaly could be that the more competitive and communicative East African groups may well have been union members in the past but found the organizations to be both partial and uninterested in the plight of coloured workers. Alternatively, the greater tendency of East Africans to work in smaller factories may well have influenced the desire and need to join a union. Both of these explanations support the notion of the East Africans as progressive but marginal elements within the Asian minority.

The data on friendships with British people again reveal a range of responses. No Punjabi Sikhs thought of themselves as being friendless within British society whereas a full 42 per cent of Urban Indian Gujuratis thought that they had no British friends.

Acculturation

The data on acculturation begin to reveal a different pattern. There is still a range of responses but this range is sharply reduced. Excluding the Sikhs and the East African Urdu speakers, most groups had a clear majority of women who still wore traditional clothing. East African groups had a smaller majority but the remainder produced values of 80–90 per cent, indicating an insignificant amount of acculturation. The data on religious adherence underscore this. No one had substituted Christianity for his own intrinsic cultural trait. Clearly acculturation is still largely superficial even amongst those groups who have adopted some Western symbols. For most groups, then, primary-group contact is still the limit of their assimilation.

Identification

No group identified with British people as their first choice, and one group even placed the British fourth in their preference list out of six. If social distance is an accurate measure of identification, it would seem that none of the Asian groups in Blackburn identifies with British society in preference to his own national group. Some, however, perceive themselves as closer to British people than others. Paradoxically, correlation analysis shows that those who did think of themselves as close to the British were also more critical of the attitudes of British people. In short, the activities of the encapsulated groups were such that they did not challenge the view that British people were essentially friendly, whilst those who competed soon came to realize that this was not the case.

Amalgamation

The marriage data confirm that none of the ethno-linguistic groups has yet progressed through to total assimilation. The highest rate of out-marriage is only 5 per cent (the East African Urdu speakers and Urban Punjabi Muslims), whilst six of the twelve groups recorded a complete absence of exogamy. Four groups recorded rates of less than 1 per cent and the standard deviation for the entire data set was only 1.75 per cent.

Synthesis

The data and analysis demonstrate a number of points. Firstly, there is a range of assimilation levels within Blackburn's Asian minority and a typology of ethnic association can be constructed from this. Secondly, the East African groups evidence greater potential for assimilation and seem to have realized this potential to a greater extent than most of their South Asian counterparts. Thirdly, urban groups appear to have become more assimilated than rural groups when the data allow such distinctions to be made. Fourthly, although there may well be a range in the potential for assimilation, and the realization of this potential may also differ, the analysis of absolute levels of assimilation reveals that none of the Blackburn groups has yet reached the position of total assimilation. Further, few groups seem to have gone beyond the stage of primary-group contact and acculturation is still largely superficial.

Having analysed the level of assimilation variable by variable, it remains to take the alternative approach of drawing together the findings into ethno-linguistic group profiles. These are contained in Appendix III at the rear of the book. Groups are listed in order of their degree of relative assimilation, based upon an average of the ranks contained in table 10.3. The Punjabi Sikhs are thus the group with the greatest degree of relative assimilation in Blackburn, whilst the Urban Indian Gujurati Muslims have the lowest. The data merit a number of comments.

First, the order in which groups appear in Appendix III could not have been predicted from the five variables designed to measure each of the groups' potential for assimilation. In other words, certain groups showed considerable potential for assimilation but never realized this. Others had little potential but overcame these barriers. In fact, a rank order correlation of the groups' potential and their realized assimilation produced a positive coefficient of only 0.23. This suggests that it is the strategies adopted by the different groups which determine whether potential is realized rather than the simple presence of innate or learned qualities. The Sikhs and Hindus exceeded their apparent potential, whilst the Indian Gujurati Muslims and the Urban group in general, did not realize their full potential.

Second, the East African groups not only possessed a good deal of potential for assimilation but also achieved this. In all cases where direct comparison was possible, East African groups had become more assimilated than their South Asian counterparts.

Third, both the Punjabi Pakistanis and the Indian Gujurati Muslims demonstrated only limited real assimilation despite having been resident in Britain for twenty years or more. They are clearly encapsulated and keen to maintain their traditional ethnic associations.

Finally, there is a general tendency for most East African groups to achieve higher average ranks for the first three phases of the model (secondary contact, primary contact, acculturation) than for the last phase (shared identity and exogamy). This may indicate that these groups are entering nascent marginality as a result of their more competitive stance.

Several of these profiles are in agreement with previously published work. All published surveys indicate that the Sikhs are a group which may lack overt potential for assimilation but which nevertheless strives for economic success and is willing to alter its behaviour to achieve this. The Pakistani Muslims seem to be a group with little potential for assimilation (or perhaps little desire) and one which has remained encapsulated in both its social and commercial inter-course. The Gujurati Hindus seem to be ill-equipped for assimilation but, despite this, they have had success in establishing secondary and primary group contacts, and they identify with British society more than most. Finally, the 'East Africans' are the least traditional of all the groups, enjoying higher levels of potential and actual assimilation than most of their South Asian counterparts. Their levels of identification do not always parallel this.

Summary

This chapter has studied the internal differences within the Asian population, especially the degree to which they might be considered assimilated. To this end, a model of the assimilation process was developed in an abstract way and only later were data mapped into this. The analysis generated assimilation profiles for each of the twelve ethno-linguistic groups which are represented in Blackburn. These profiles can be ranged along a notional continuum of ethnic association such that assimilated groups lie at one end and encapsulated or traditional groups lie at the other. The Blackburn data indicate that the East African groups occupy the more assimilated positions on the continuum but that, even here, assimilation is not total. This reinforces the conclusion that the East Africans are likely to engage in usurpationary closure in order to gain parity with the white working and middle classes. The hostility of white society, and its desire to enforce exclusionary closure seem likely to leave the East Africans in a marginal and vulnerable position which they will resent. The assimilation model suggests that, as a result, these groups will not pass beyond the acculturation stage. Conversely, those groups which occupy positions towards the unassimilated end of the continuum appear to value success at encapsulatory closure more highly than they do usurpationary closure. This scale of priorities results from the groups' beliefs in migration as a temporary phenomenon, and it will prevent them passing beyond the primary-group contact stage.

As a secondary consideration, this chapter has also been concerned with relations between the different Asian sub-groups. The data on residence and intermarriage seem to argue persuasively that members of many of the groups have as great an interest in maintaining social distance from what they consider to be less desirable Asian sub-groups as they do in maintaining distance from whites. The indices even suggest that some groups value the avoidance of contact with other Asians *more* highly that they do avoiding contact with whites.

The size of the Blackburn samples encourages these conclusions to be presented as being of local validity alone. However, this chapter does indicate an area of research which has, to date, been significantly under-researched. Although the literature is therefore still only nascent, there are already common themes and findings to which the Blackburn analysis makes a contribution.

CHAPTER 11

Summary and Conclusions

The General Approach

The focus of this book has been the migration of Asians to Britain, and in particular the movement of that group to the northern industrial town of Blackburn. However, before any conclusions are presented and even before the argument is summarized, it is essential to make one point. This is that, because of the nature of Blackburn and the characteristics of its Asian inhabitants, this case study cannot be regarded as one which is necessarily typical. Blackburn is not a large conurbation, it does not suffer from the acute housing shortages which afflict many inner-city areas, its immigrant population is largely from one racial group, and its Asian residents appear to be more conservative in their ways than those found in other cities. In short, the findings of this study cannot be generalized to cover all areas of Asian settlement, let alone all areas of immigrant settlement. Having said this, the same is true of the majority of the community studies which have been published over the last twenty years. Each relates only to its own locale and each reflects the impact of differing local histories, local opportunities, and local patterns of immigrant settlement. As was argued in chapter 1, this plethora of local studies has by no means been lacking in value, and such empiricism is an essential pre-requisite both of the growth of a sub-discipline and of the shift towards classification and explanation. At one level, then, this book can simply be taken as a further contribution to the burgeoning literature on local community studies. It describes which groups have settled in Blackburn, the opportunities presented by the local labour and housing markets, and the reaction of the white population to this in-migration. It also describes, at a variety of scales, the patterns of social and spatial behaviour produced by the Asian population. At this level the book is thus a study of a unique geographical region and its unique population.

However it is intended that the book can also be viewed at an entirely different level if the reader so wishes. It is an attempt to explain how the coloured population of Britain fits into that country's social, economic, and residential structure. In this context, Blackburn simply becomes a part of the real world which is mapped into the theoretical explanation in order to test various key hypotheses arising out of the model presented in chapter 6. That Blackburn was chosen along with its Asian population becomes almost irrelevant. It represents a convenient natural laboratory but one of many such examples which exist throughout Britain. What is of importance in this context is that Blackburn allows us to see how white British society operates, how the different classes and status groups engage in different forms of social closure, and how different immigrant groups have contrasting aspirations and potential that are reflected in the strategies which they adopt whilst in Britain. It demonstrates how white society carefully circumscribes the limits of action within which immigrant groups are allowed to exist; how these limits are operationalized through exclusionary closure and through gatekeepers; how certain immigrant groups who seek social mobility through market competition are forcefully reminded of these limits by persistent discrimination and rejection, and therefore why these groups become marginal, frustrated, and resentful; and how other immigrant groups who determine their strategies and actions by reference to the sending society and its values, may rarely become completely aware of the barriers which circumscribe them because of their encapsulation. At this level, then, the book is a theoretical comment on British society, and the relationship of that society with its immigrant populations. The specific empirical support for these contentions could have been derived from a study of the Jews in South Wales, the Poles in London, the Chinese in Liverpool, or the Vietnamese refugees in Worcestershire.

These two approaches are not mutually exclusive but they often fail to be complementary. A conscious attempt has been made in this book to ensure that the two do not stand in isolation but are integrated into a unified whole. The keys to this are chapters 2, 3, and 10. Chapter 2 demonstrated that whilst each city or town has elements of uniqueness which could justify a particularistic case-study of that settlement, in practice many urban areas have features and characteristics in common with others and can therefore be grouped into types. This is not to deny the importance of unique local history or conditions but simply to note that in many cases, economic and social similarities override the purely local differences. In this way, settlements can be classified and national typologies can be created. Once this is achieved, local case studies take on a different meaning. They are no longer descriptions of unique urban regions (i.e. Blackburn) but studies of representative examples of a national type of settlement (i.e. textile towns). The conclusions derived from one such study can be transposed, with care, to apply to a whole series of economically and socially similar settlements elsewhere in the country. In this way research can be channelled into those urban categories which have previously been underresearched and the duplication of parallel research effort prevented. Chapter 2 consequently linked the unique with the general, and the empirical with the theoretical. Chapter 3 had a very similar function but, whereas chapter 2 was concerned with placing Blackburn in a synchronic typology of urban centres, chapter 3 aimed to locate the ethnic settlement in Blackburn in a temporal or diachronic context. It was therefore concerned with describing how the pattern and form of ethnic settlement in British cities has changed through various phases from the pioneer stage of the 1950s to the suburbanization and municipalization of the 1980s. Again, a typology of this nature allows the case study of Blackburn in the late 1970s to be grouped with other studies of this particular temporal phase of ethnic settlement. Similarly, it allows this book to be separated from say Rex and Moore's (1967) study of the lodging-house era or Little's (1948) study of the early pioneers. Again, the unique can be associated with the general. Finally, chapter 10 presented an assimilation typology based upon the social, economic, and psychological characteristics of members of a

given migrant group. This typology again provides a context within which the
Blackburn Asian groups can be located and which allows classification of
similar and dissimilar ethnic or ethno-linguistic groups. Encapsulated South
Asian minorities might therefore be grouped with encapsulated Chinese
minorities, and yet again the particular can be seen in relation to the general.

The way in which these three typologies coincide to locate the Blackburn
study, not as a unique entity, but as only one example of a particular combination of time, spatial distance, and social distance, is illustrated graphically in
figure 11.1. It is in this way that the book can be both a description of unique
local phenomena and a theoretical analysis of generalizable circumstances.
Clearly the next step is progressively to broaden the scale of analysis to see how
different combinations of time, space, and social characteristics require or
generate different explanations to those proposed in chapter 6.

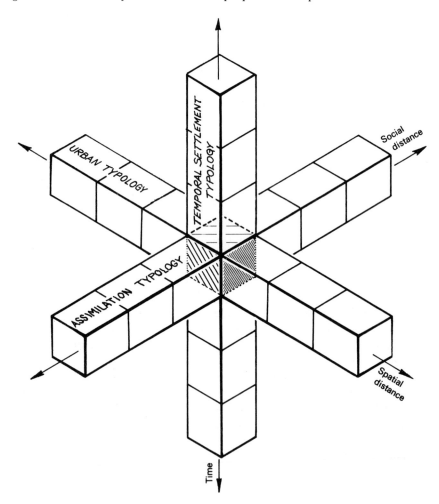

FIG 11.1: The context of the Blackburn study

Findings

Although the findings of each individual chapter have been presented therein, they are here drawn together to provide an integrated summary.

Chapters 4 and 5 were concerned with identifying and enumerating the various forces which are likely to bear upon the behaviour of Asians in Britain. Chapter 4 looked at the form of British society, the attitudes of its members towards coloured immigration, and the extent to which these attitudes were operationalized as behaviour. However it was acknowledged that the attitudes of an individual, let alone those of a group, are multi-faceted and that it is impossible to either describe or analyse every element of them. Instead attitudes towards only one interface between immigrants and core population were selected for further study. This was the issue of housing. The remainder of the chapter therefore showed the reaction of the core population to coloured immigration through the example of how that population manipulated the housing market. Incontrovertible evidence exists, both at a national and local level, to show that white society persistently and systematically places barriers in the way of those coloured immigrants who wish to gain access to better housing in better residential neighbourhoods. These barriers are often not impermeable, but taken as a whole they effectively relegate the coloured population into a separate, excluded underclass. Moreover, evidence not cited in this book suggests that this situation is not unique to the housing market and that the majority of scarce resources are allocated in such a way that the coloured population is disadvantaged and debarred. Whilst it is possible to argue that it is not abnormal for a group to protect its own interests at the expense of 'outsiders', and that this exclusion takes place irrespective of the characteristics of newcomers, the scale, intensity, and persistence of exclusion against blacks suggests that such an argument is untenable in this case.

Chapter 5 focused upon the other end of the migration stream and the way in which social and historical factors interacted to implant certain attitudes and behavioural norms in migrants prior to their arrival in Britain. These were again thought to be factors which could bear heavily upon patterns of Asian behaviour in this country. In particular, the chapter commented upon the importance of the joint family, respect for elders, and the enhancement of social status. This last issue was looked at in greater detail, and it was concluded that, because British colonial rule had significantly raised aspirations but had done little to improve tangible opportunities, the Asian population had resorted initially to rural–urban migration but latterly to international migration to Britain. Migrants had, however, arrived in Britain with 'attitudinal baggage' which incorporated a negative evaluation of many aspects of British society and life, and a strong belief in a myth of return. The encapsulation which resulted from these attitudes ensured that stereotypes and opinions were never tested or challenged. Continued physical, financial, and emotional links back to the sending society underpinned modes of behaviour and thinking. In short, although many groups would never return *en bloc* to their sending societies, most South Asian minorities continued to behave as if they would, and continued to act out their lives according to a code of practice based upon traditional Indian or Pakistani principles.

Having described and analysed every aspect of the context of Asian migration to Blackburn, from the physical conditions in the sending and receiving societies to their social conditions, it seemed appropriate to integrate this into an holistic model which would suggest key areas or hypotheses which could be investigated empirically. It was argued, however, that the concepts which had been introduced into the mainstream of the literature on Asians in Britain were both incomplete and inadequately developed. As a result, several other theoretical constructs were advanced or expanded in order to shed additional light on the behaviour of Asians in Britain, and on their position in the social structure. Parkin's ideas on 'social closure' allow an accurate description not only of relations between the classes within British society, but also of how whites react to the presence of coloured people, and how the latter might attempt to gain full entry to British society. Mayer's notion of 'encapsulation' seems to be particularly appropriate to the South Asian population in Britain, and the concept of 'marginality' appears to have *a priori* relevance to East Africans.

Chapter 7 was the first empirical chapter. It was concerned with the migration to Blackburn and therefore with a micro-scale example of those general trends outlined in chapter 3. In particular, it sought to differentiate between the two major migration streams, notably that which brought Asians direct from the sub-continent and that which involved Asians from East Africa. The former were found to be economic target migrants who came to Britain during the labour shortages of the late 1950s and early 1960s. They possessed a sojourner mentality predicated upon a belief in the myth of return. Their backgrounds prior to migration acted to underscore these tendencies, since the South Asians in Blackburn were largely from rural, agricultural backgrounds where they had received little formal education and no tuition in English. Consequently, even if the South Asians had wished to participate widely in British society, which they did not, they were singularly ill-equipped to do so. In contrast, the East Africans were involuntary political refugees who had entered Britain in the 1970s at a time when the national economy was already showing signs of recession. The East Africans did, however, come from backgrounds which are likely to encourage them to re-establish their businesses and middle-class life-style through competition and usurpationary closure. Chapter 7 therefore suggested that South Asians and East African⋅Asians might be expected to adopt different roles in British society and different attitudes towards settlement here.

Chapter 8 ignored the differences which exist within the Asian minority in Blackburn and instead looked at the external relations of that minority with the local white·population. Clear evidence was presented to show that whites engage in exclusionary closure in all the major areas which determine social status. The latter part of the chapter then looked at how the Asian population reacts to this situation. In general it was found that they do not enter into usurpationary closure in order to gain fuller access to society, and that many of

the scarce resources which are protected so jealously by whites are in fact not
desired by Asians. The reason behind this is that Asians possess a very different
value-system from local whites and that, for many, this system is firmly orientated around encapsulatory closure prior to return migration. Behaviour is
therefore guided by the reference group of the sending and not the receiving
society.

Chapter 9 returned to the theme of chapter 7, namely that 'the Asian'
population contains South Asians and East African Asians and that each of
these groups is likely to adopt a different role in British society. Chapter 9
determined whether this was true in Blackburn. It was argued that it was South
Asians who were found to have more restricted aspirations whilst in the UK
and their orientation seemed to be characterized by frugality and the myth of
an early return to their homeland. Although they suffered conditions which
were objectively inferior, their limited contact with individuals who held alternative standards and their continual interaction with like-minded fellow
migrants ensured a situation where South Asians were less conscious of their
relative deprivation. Furthermore, because of their introspection and orientation towards their homelands, few South Asians have fully entered into any
form of usurpationary closure against white society. Consequently, in many
respects they remain unaware of the extent to which exclusionary closure is
practised by whites. In addition, relations *within* the South Asian group were
also explored and it was noted that groups not only engaged in encapsulatory
closure against whites but also against other dissimilar South Asians. Again,
the myth of return, the desire to retain cultural purity, and the persistence of
pre-migration prejudices and preferences go a long way towards explaining
this phenomenon.

The conclusions for the, admittedly conservative, East Africans were
different. Their aspirations were less restrictive, they retained a healthy drive
to social and economic mobility within whatever societal structure they found
themselves and they appeared to exhibit a more limited desire to retain
community cohesion. They were also found to be more willing, and better able,
to engage in usurpationary closure against whites. However, such willingness
to discard ethnic defences results in a wider exposure to the exclusionary
constraints erected by local whites. Ethnic exclusion by the white reference
group can lead only to marginality for the East Africans.

Chapter 10 altered the scale of analysis yet again and focused upon the
differences between the various ethno-linguistic groups within the Asian population. Twelve of these groups were identified as being sufficiently large in
Blackburn to allow detailed analysis, although it was admitted that the small
size of several of these meant that conclusions could only be tentative. Prior to
the presentation of data, a model of ethnic association was outlined based
loosely upon Gordon's 1964 assimilation chronology. This model proposed a
series of assimilation phases through which groups might pass on their way to
'total assimilation', and also indicated the key barriers which might prevent
progress from one phase to another. Variables were then selected either as
indicants of the phase of assimilation or as surrogates for the barriers. The

analysis of these data demonstrated that none of the groups in Blackburn was totally assimilated, and more than this, that no one group was even near to this state. However, the groups could be arranged along a continuum of ethnic association and their level of relative assimilation identified. The results indicated that the more assimilated groups tended to be East Africans, whilst the South Asians remained intent upon encapsulatory closure.

Conclusions

Whilst the single most important and valuable point to come from this study is that 'the Asian' population is in fact a series of independent and different sub-communities, several conclusions have been suggested which related to that population as a whole. It has been argued forcibly that, regardless of whether one looks purely at the issue of housing or takes a more general view, Asians are a deliberately excluded and disadvantaged group. In this respect Rex's conceptualization of Asians and other coloured groups in Britain as an underclass is fully supported. Moreover the account presented here also agrees with Rex that the motives behind exclusionary closure by whites not only include simple economic ones but also the fear of loss of social status. The roots of this fear can be traced back to the colonial era and to the social and power relations which existed between European colonizers and Asian subjects. As Cox (1948) argued, it was essential that thinking Christians rationalized their economic exploitation of colonial labour forces and raw materials through the development of derogatory stereotypes which cast the indigenous populace in the role of inferiors. The white man's burden. Only if the 'native' population could be thought of as inferior could their inferior treatment be justified. Such attitudes were carried back to metropolitan society where they became entrenched in the educational system and the national psyche. However none of this made allowance for the fact that one day these 'inferiors' might migrate to Britain, settle in British cities, live in British neighbourhoods, and work in British factories. When that occurred, the key spatial and social distance which had been essential for the operation of colonial exploitation in the non-metropolitan context no longer existed. British people who had been taught for years that Asians were inferiors suddenly found that members of that group had moved into their street, bought houses like theirs, acquired jobs as good as theirs, and were now sending their children to the same schools as theirs. Status anxiety was inevitable under these circumstances, especially given the Asians' willingness to work for lower wages, their strong culture, their different language and religion, and their quite natural desire for group cohesion and encapsulation. The onset of economic recession added the final seal.

It is clear that the continued relegation of Britain's black population to an underclass necessitates renewed efforts to overcome racial disadvantage, prejudice, and discrimination. British society is by no means as colour blind as either the public or successive governments like to think. That both the population and administration are allowed to continue to delude themselves about the plight of Britain's black population is clearly the fault of the Commission for

Racial Equality (nee Community Relations Commission). It is that body which ought to be persistently reminding us of the ethnic disadvantage which occurs in Britain, in order that a climate of opinion is created within which change can occur. A non-discrimination notice merely victimizes one firm, one local authority, or one individual whilst the rest of society is left untouched. It tries to stop the problem of smoking by destroying one ashtray. What is needed is not more isolated and lengthy court cases but an attempt to change social attitudes; to make it socially and morally unacceptable for 35 per cent of people to admit to being prejudiced (Airey, 1984); to make it impossible for British people to condemn apartheid in South Africa whilst remaining unconcerned about the blatant and ubiquitous discrimination which takes place within their own country. Such a wholesale change in attitudes will not take place overnight but other examples such as the way in which attitudes towards nuclear disarmament, nuclear power, or even smoking have been manipulated and changed by careful campaigns show what can be achieved in the long term. The riots of 1980 and 1981 and Scarman's (1981) incriminating commentary on British race relations show that such actions need to be taken sooner rather than later. Moreover, they need to be part of a coherent and continuing strategy towards the issue of race rather than isolated incidents within a sporadic and piecemeal campaign motivated and propelled by successive crises. The legislation on immigration and the way in which Britain copes with refugee resettlement (Robinson, 1985b) exemplify this latter trend par excellence. Finally, on this issue, one should perhaps point out that criticism of Britain's record on race relations and societal attitudes towards coloured people does not arise purely within guilt-ridden liberal academic circles. Even brief acquaintance with Asians who have experience of the way in which other administrations and societies handle race relations shows that they are highly critical of Britain and her attitudes. Invidious comparisons are often made between life and attitudes elsewhere and those in Britain.

However, whilst much of this book has taken a similar line to the current 'conventional wisdom' in British race relations (exemplified by Rex), other parts challenge that wisdom. The majority of coloured people in Britain may well be objectively a part of a single underclass (or housing class) but this does not necessarily produce cohesion or even commonality of interest and identity. Whilst these various classes may be classes-in-themselves, they do not become classes-for-themselves unless one is prepared to lend extraordinary emphasis to multi-ethnic political groups centred on Marxist philosophies. In other words, whilst Rex and others may be correct about the role and form of exclusionary closure, their analysis of the impact of this on patterns of behaviour and self-identity is flawed. Notions such as underclasses and housing classes are valuable up to a point but they need to be complemented by fuller analysis of those ethnic minorities themselves, their aspirations, expectations, and potentialities. Only when these areas have been exhaustively researched (especially by continual reference to the sending societies and their social and environmental conditions) can we hope to *understand* fully the behaviour of the different ethnic groups as represented in Britain. We cannot naively assume common aspirations, drives, and standards even within the coloured population let alone within our multi-racial society as a whole.

It is this inadequacy in current thinking which prompted the structure of this book. The macro-level structural forces which delimit the outer bounds of permissible action have been described and analysed in detail but these have been set against those group-specific cultural forces which determine whether such bounds are ever transgressed or even approached. In looking at the dialectical relationship between exclusionary barriers and the motivations of the encompassed Asian population, one major distinction has been drawn and many others highlighted. It has been shown that, for South Asians both as a minority and as separate groups, the latter set of forces is perhaps of greater immediate importance in explaining behaviour than is the former. The desire for encapsulatory closure, legitimized by the belief in return migration, seems to be a more potent force in determining the bounds of South Asian mobility and behaviour than do the encompassing constraints. This finding challenges the current assumption in academic and government circles that assimilation or integration is equally *desired* by all the different coloured groups in British society. Moreover it also challenges the assumption that policies designed to enforce integration are equally *desirable* for all groups.

The issue of possible voluntary South Asian return migration to the sending society has featured strongly in the book. It is important to stress, however, that under no circumstances should such possibilities be used to support or validate any policy of repatriation, or any policy which seeks to deny equal treatment to any group whilst they remain in Britain. The evidence presented here makes it clear that the myth of return is, as yet, only a way of thinking for the majority of Asians and not a practical short-term likelihood. This is not to deny the importance of this belief since it acts as a significant and long-term constraint upon the behaviour of many South Asians in Britain.

The same holistic approach to structural and cultural forces was also used in relation to the East African Asians, again both as one minority and as separate groups. It was argued that they have faced similar structural barriers to their South Asian counterparts but that they have reacted, and are likely to react, to these in a very different way. The different potentialities which their pre-migration experience afforded them, the contrasting reason for their presence in Britain, and their general outlook indicate that this group have already become much more aware of the exclusionary barriers which prevent their social mobility, and that they have consequently resorted to usurpationary closure in an effort to gain equal treatment. It was also argued that their internalization of many western values and their exclusion from entry to British society was likely to leave them in a marginal social position. It will be extremely valuable, and of great potential significance for all Asians in Britain, to see how this group responds to such a position. The Blackburn data, and the timing of the two surveys prevent any prediction about whether they will articulate their challenge through conventional channels (such as politics), through alienation, or through re-emigration.

Overall, the conclusions of this book are that, whilst events for the West

Indian population may have already developed beyond the point where remedial action could be sure of success, this is not the case with the Asian populations. The myth of return amongst South Asians, and the refugee status of East African Asians, have given British society an unprecedented breathing space in which it can demonstrate, through comprehensive and wide-ranging action, that it is no longer a racist or discriminatory society and that it does not see these groups only as an obsolete input to processes of industrial production. It is to be hoped that this opportunity is used wisely.

A Description of the Nine Most Common Clusters
Found in Blackburn; 1971

Cluster 23 consists of poor quality housing which is seriously lacking in amenities. Owner-occupation and unfurnished renting are common but overcrowding is below the national average. Heads of household are largely unskilled or skilled manual, and employment is mainly in manufacturing. The population is biased towards the older age groups and in-migration is not common. Other demographic characteristics follow the national average. This cluster contains 12 per cent of Blackburn's population.

Cluster 52 contains poorer housing than cluster 23. Household amenities are even less common. The presence of New Commonwealth immigrants is marked, as are over-crowding and private renting. The population is largely semi- or unskilled and is employed in manufacturing industry. The age structure is weighted towards the younger cohorts, and single non-pensioner households exceed the national average. Unemployment is also relatively common. This cluster contains 10 per cent of the town's population.

Cluster 22 also contains very poor quality housing of the terraced variety. The provision of amenities is similar to cluster 23. However, 22 differs in several important respects; the female activity rate is higher; the areas have excellent accessibility to local employment; the employment structure is skewed more towards semi-skilled heads of household, and non-manual occupations are more common; 22 had a much higher rate of furnished tenancies than 23; and unfurnished renting is also more frequent—as a result, overcrowding is more a problem although still below the national average; and single non-pensioner households are more in evidence. Approximately 9 per cent of the population live in such EDs.

Cluster 21 consists of the poorest older terraced housing which has not seen ethnic in-migration. Half of households rent their accommodation in an unfurnished condition from private landlords, and two-thirds lack the use of an inside WC. The cluster evidences a very high proportion of unskilled and unemployed workers, and is attractive to young families with young children. Seven per cent of Blackburn's population reside in such areas.

Cluster 33, despite being in family five (areas of extensive public housing), is predominantly owner-occupied. The CES include it in this family since one would normally expect the residents to become local authority tenants. The cluster does have a higher level of public renting than all the others which appear in Blackburn except one. The population is largely employed in manufacturing but is skilled or semi-skilled. There is a high female activity rate and low unemployment. Housing is mostly low-value inter-war developments which evidence below average overcrowding. This cluster, which contains 6 per cent of the population, is characterized by low levels of social deprivation.

Cluster 12 includes areas of high status. The workforce is largely professional or managerial and a smaller percentage of women work. Car ownership is high. Housing can be described as being inter-war suburban of above average size and quality. Owner-occupation is prevalent and residential turnover is low. The population is elderly. Such areas house nearly 6 per cent of the total population.

Cluster 25 contains areas of older terraced housing which are in good condition and owner-occupied. Overcrowding is below average. The cluster is distinctive for its very high proportion of skilled manual households employed in manufacturing, its low female activity rate and low levels of service employment. The population is stable and biased towards the older age ranges. Cluster 25 contains 5 per cent of the population.

Cluster 28 has even better housing than cluster 25. Housing is slightly newer and better provided with amenities. Owner occupation is even more entrenched. The population is again elderly and enjoys an above average socio-economic status. Cluster 28 also houses 5 per cent of the population.

Cluster 27 is similar in its housing conditions to 25 but with greater access to amenities and a lower level of owner occupation. Council tenancy replaces this form of tenure. A further distinguishing feature is the predominance of semi-skilled heads of household. These areas contain 4 per cent of the town's population.

APPENDIX II

The Social Surveys

Unless it has been specifically stated otherwise the data analysed in this book have been derived from one of the two social surveys undertaken in Blackburn during 1977 and 1978. The first of these has been referred to throughout as the 1977 Asian Census. It sought specifically to provide a mid-term census of Asian households to allow conclusions to be based on more contemporary information than was available from the national census of 1971. The questionnaire consequently contained only those questions needed to derive social and demographic data and was designed intentionally to minimize the length of the interview and therefore maximize the response rate. Interviews usually lasted less than five minutes and were normally conducted on the respondent's doorstep. In aiming for as large a sample as possible the sampling frame which was chosen as the most effective was the electoral roll. In view of the acknowledged under-registration of those Asians eligible to vote, the sample was extended by snowball interviewing in which respondents were asked to notify interviewers of Asians whom they knew, but were known not to be on the electoral roll. Ninety-seven additional Asian families were contacted in this way. However, of the 2204 Asian households who were thus included in the sampling frame, 95 proved untraceable (after three return visits at different times of the day and on different days of the week) and 407 refused to be interviewed. Ultimately, therefore, data were collected for 1702 households, or 77.2 per cent of those approached. Where calculations required only the number of Asians in a street, enumeration district, or ward, those families who were contacted but refused to be interviewed have been included. In other cases, no attempt has been made to adjust conclusions to make allowances for non-respondents. The Asian Census was completed by the end of 1977 and was successful in collecting data on 4721 adults and 4141 children.

The 1978 Sample Survey was designed to achieve different objectives and sought to gather in-depth quantitative and qualitative data from a much smaller number of households. The sampling fraction was not fixed in advance but was determined instead by the available time. The sampling frame which was employed was the 1978 Electoral Roll, in order that a realistic sub-sample of those Asians who had moved into council housing could be assembled. Use of the 1977 Electoral Register would have made this more difficult. The Asian population was then stratified by area of residence (as defined by the cluster analysis of 1971 Census data) such that seven of the eight possible 'families' were represented. The sampling fraction within these strata was not constant but was varied to increase coverage in peripheral higher-status areas (to 100 per cent) and depress coverage in the major core areas (to 15 per cent). Having said that, the residential and social structure of Blackburn is such that most interviewees were still residents of core areas. Within these strata, households were selected for interviewing by simple random sampling. As with the 1977 Survey, households were notified by letter in advance that they would be visited. The letters also explained the aims of the surveys and gave a contact point where enquiries could be made if respondents felt worried about any aspects of the survey. In the case of the 1978 Survey, each household was then visited in order to arrange an appointment at the interviewee's convenience. Households were contacted up to five times before being classified as non-respondents.

For the 1977 Survey, prior knowledge of the town allowed careful matching of inter-viewers with the regional-linguistic characteristics of those Asians known to live in a particular area. For the 1978 Survey, the accuracy of interviewer-interviewee matching could be increased by making use of preliminary analysis of the 1977 Asian Census. In 1978, interviewers worked in pairs (male and female) to allow simultaneous and more effective interviewing of both husband and wife. If either of these was absent when the interviewers called, a return visit could be made later to see him or her separately. The interview schedule for the Sample Survey contained 185 questions, of which 138 were pre-coded, the remainder being structured discussions. The emphasis during inter-viewing was firmly upon the quality of data collected rather than its quantity. Inter-viewees were allowed, or even encouraged, to digress if it appeared that they wished to discuss other issues which were relevant or interesting. The interview rarely took less than one hour and frequently extended to as long as 90 minutes. By the close of the data collection phase of the survey, 500 households had been contacted and 391 part, or whole, households had been interviewed. The response rate was thus 78.2 per cent, and the total sample represented around 18 per cent of all Asian households on the Electoral Roll. The specific wording of questions that were asked in the 1977 Asian Census and 1978 Sample Survey are provided within the main body of the text wherever the results are presented.

APPENDIX III

Assimilation Profiles of Twelve Ethno-Linguistic Groups in Blackburn

1. *Punjabi Sikhs* evidenced little potential for assimilation but in practice were the most assimilated group. They produced high rank scores on all variables except exogamy. However, whilst in relative terms they were structurally integrated, acculturated, and shared a common identity with the core culture, in absolute terms they were not totally assimilated. The group did not out-marry and over 50 per cent of them would have to change their ward of residence to replicate the same spatial distribution as that of the white population. Clearly, then, whilst they were the most assimilated of the Blackburn groups, they had still not become totally assimilated.

2. *East African Urdu-speaking Muslims* were consistently assimilated on almost all criteria although they did not regard themselves as being socially close to the British. They also demonstrated high potential for assimilation.

3. *East African Gujurati Hindus* had only moderate potential for assimilation but had succeeded in becoming one of the more assimilated groups. They were relatively acculturated, identified with British society, and were keen union members. Their structural integration and levels of exogamy were lower.

4. *Rural East African Gujurati Muslims* had good potential for assimilation and had succeeded in realizing this. They were structurally integrated (although their union membership was low), acculturated, and they were moving towards an identity shared with the core culture. There was, however, no exogamy.

5. *Indian Gujurati Hindus* showed little potential for assimilation. They had a skewed social-class distribution, few linguistic skills, and only a limited orientation towards permanent settlement in the UK. Despite this, they had achieved considerable relative assimilation. In particular they scored highly for their union membership, primary-group contact, and paradoxically, their perception that British people were friendly and not dissimilar to themselves.

6. *Urban East African Gujurati Muslims* displayed considerable potential for assimilation. This was translated into relatively strong structural integration, a willingness to identify with British society, and some out-marriage. Levels of acculturation were relatively low.

7. *Indian Urdu-speaking Muslims* possessed moderate potential for assimilation and this was, in fact, the degree of relative assimilation which they had achieved. They were characterized by contrasting patterns of secondary-group contact (strong residential but weak industrial integration), poor relative identification with British society, and an absence of exogamy. Set against this was their moderate acculturation.

8. *Pakistani Urdu-speaking Muslims* evidenced only limited potential. In practice, they were weakly assimilated although their degree of secondary integration and identificational assimilation were moderate.

9. *Rural Pakistani Punjabi Muslims* were also weak on potential for assimilation. However they did produce moderate to high rankings on exogamy and identification although

their secondary integration and acculturation were both very weak.

10. *Rural Indian Gujurati Muslims* were less weak on potential than either of the two previous groups. Although the extent of their residential segregation was relatively low and their levels of out-marriage quite high, other scores were below average. The group could not be described as structurally integrated and did not identify with British people.

11. *Urban Pakistani Punjabi Muslims* exhibited a good deal of assimilation potential. Their social class profile was the least skewed of all the twelve groups and they were relatively competent at English. Despite this, the group has realised little of this potential except in the sphere of exogamy. They did not appear to identify with British people and their acculturation was also limited.

12. *Urban Indian Gujurati Muslims* evidenced moderate potential for assimilation but had not achieved this. They were the most weakly structurally integrated of all the groups and their level of acculturation was also low. Identification was also weak.

TABLES

TABLE 2.1. *The twenty largest ethnic settlements in 1971 and 1981*

Rank	Coloured population 1971	NCWP population 1981[1]
1	Greater London	Greater London
2	Birmingham	Birmingham
3	Wolverhampton	Leicester
4	Leicester	Bradford
5	Bradford	Wolverhampton
6	Manchester	Manchester
7	Coventry	Warley and West Bromwich
8	Leeds	Coventry
9	Nottingham	Huddersfield
10	Warley and West Bromwich	Leeds
11	Huddersfield	Luton
12	Walsall	Slough
13	Luton	Nottingham
14	Sheffield	Walsall
15	Derby	Bolton
16	Slough	Derby
17	Bristol	Sheffield
18	Bolton	Bristol
19	Reading	Blackburn
20	Blackburn	Oldham

Note: [1] As mentioned in the text, these data are not directly comparable with those for 1971

TABLE 2.2.   *Moser and Scott's urban typology*

Type	Group[1]	Example
1. Resorts, administrative, and commercial towns	Seaside resorts	Worthing
	Spas, professional and administration	Bath
	Commercial centres	Southampton
2. Mainly industrial towns	Railway centres	Crewe
	Large ports	Birkenhead
	Textile centres	Huddersfield
	North-east seaboard and mining	Gateshead
	Recent metal manufacturing	Scunthorpe
3. Suburbs and suburban towns	Exclusive suburbs	Esher
	Older mixed-use suburbs	Ealing
	Newer mixed-use suburbs	Chigwell
	Light industry suburbs	Gosport
	Older working-class suburbs	Willesden

Note: [1] These are abbreviated titles. For full description see p. 17 of Moser and Scott.

TABLE 2.3. *Armen's urban typology*

Type of centre	Class	Sub-class	Example
1. Conurbations	Inland		West Midlands
	New		Gtr. Sheffield
	Maritime		S.E. Lancashire
2. Service	Port	Merseyside	Merseyside
		Large port	Teesside
		Medium port	Gtr. Swansea
		Small port/resort	Gtr. Ipswich
		Day trip/vacation	Gtr. Portsmouth
	Holiday/retirement	Small port/resort	Gtr. Ipswich
		Day trip/vacation	Gtr. Portsmouth
		Vacation/retirement	Morecambe
		Coastal retirement	Southport
		Inland retirement	Windsor
		Admin./defence	Gtr. Cheltenham
	Admin./Market	Admin./defence	Gtr. Cheltenham
		Market/cathedral	Gtr. Exeter
		Admin./court	Gtr. Lincoln
		University/market	Gtr. Bath
		Oxbridge	Oxford
3. Manufacturing	Industrial	Railway	Gtr. Preston
		Engineering	Gtr. Luton
		Traditional	Gtr. Burnley
		Specialized	The Potteries
		Expanding manuf.	High Wycombe
	Rapidly developing	Expanding manuf.	High Wycombe
		Industrial estates	Kirkby
		Maturing new towns	Basildon

TABLE 2.4.   *Webber and Craig's urban typology*

Family	Cluster[1]	Example[2]
1. Suburban and growth centres	High status plus manuf.	Stafford
	Rural growth	Newbury
	Rapid growth	Wycombe
	Older high status residential	Tandridge
	Large student population	Barnet
	Outer London	Hillingdon
2. Rural and resort areas	Rural Wales	Glyndwr
	Rural West	Suffolk coastal
	Rural East	Babergh
	Rural Scotland	Perth
	Resort retirement	Arun
	Port retirement	Lancaster
3. Traditional industry and mining areas	Lowland heavy industrial	Wakefield
	Upland heavy industrial	Sheffield
	Black Country	Sandwell
	Large industrial plants	Thamesdown
	Small town manufacturing	Chorley
	Pennine towns	Bolton
4. Service centres	Metropolitan	Nottingham
	East End of London	Southwark
	Scottish	Kyle
	Regional	Southampton
	Wales/Merseyside	Wirral
5. Areas with much local authority housing	Scottish industrial	Renfrew
	Overspill	Knowsley
	New towns	Harlow
	Glasgow	Glasgow
6. Inner and Central London	Kensington and Chelsea	Kensington and Chelsea
	Central London	Westminster
	Inner London	Wandsworth

Notes: [1] for full descriptions see Webber and Craig pp. 83–5
         [2] these are the towns/areas which are the statistically most representative of their cluster.

TABLE 2.5. *Donnison and Soto's urban typology*

Group	Family	Cluster	Example
Non-manual	1	London	London
	2	Regional service centres	Reading
	3	Resorts	Torbay
	4	Residential suburbs	Cheadle
		New industrial suburbs	Runcorn
		New towns	Harlow
Manual	5	Welsh mining towns	Aberdare
		Engineering I	Stockport
		Textile	Accrington
		Engineering II	Luton
		Heavy engineering/coal	Hartlepool
	6	Inner conurbations	Birkenhead
		Central Scotland	Dundee

TABLE 2.6. *A typology of ethnic settlements in 1971*

I. LONDON
   (a) Traditional industrial areas: Barking, Slough, Gravesend, Luton, Enfield
   (b) New industrial areas: Watford, Hillingdon
   (c) Inner areas: Brent, Haringey, Ealing, Lambeth, Wandsworth, Islington, Hammersmith, Westminster, Camden, Kensington and Chelsea
   (d) Suburbs: mixed residential: Croydon, Hounslow, Redbridge, Richmond, Kingston, Merton, Harrow, Bexley
                    exclusive residential: Barnet, Bromley, Sutton
   (e) Service: Lewisham, Waltham Forest, Greenwich, Hackney, Newham, Southwark, Tower Hamlets

II. CONURBATIONS
   (a) Traditional industrial areas: Birmingham
   (b) Service/port areas: Liverpool, Newcastle
   (c) Service/textile areas: Manchester, Nottingham, Leeds

III. SERVICE CENTRES
   (a) Commercial: Bristol, Reading, Southampton, Cardiff, Portsmouth
   (b) Professional and administrative: Oxford, Bedford

IV. INDUSTRIAL CENTRES
   (a) Textiles: Bradford, Bolton, Blackburn, Rochdale, Oldham, Halifax, Dewsbury, Ashton-under-Lyne, Huddersfield
        and services: Leicester, Preston
   (b) Engineering
        and commerce: Northampton, Gloucester
        and railways: Coventry, Sheffield, Derby
   (c) Heavy engineering: West Bromwich, Wolverhampton, Walsall, Dudley, Warley

TABLE 2.7.   *A typology of ethnic settlements in 1981*

I.   **LONDON**
  (a)   Traditional industrial areas: Barking, Slough, Gravesend, Luton, Enfield
  (b)   New industrial areas: Watford, Hillingdon, *Gillingham*
  (c)   Inner areas: Brent, Haringey, Ealing, Lambeth, Wandsworth, Islington, Hammersmith, Westminster, Camden, Kensington and Chelsea
  (d)   Suburbs: mixed residential: Croydon, Hounslow, Redbridge, Richmond, Kingston, Merton, Harrow, Bexley
             exclusive residential: Barnet, Bromley, Sutton, *Havering, Epsom, Esher, Reigate, Godalming, Haywards Heath*
  (e)   Service: Lewisham, Waltham Forest, Greenwich, Hackney, Newham, Southwark, Tower Hamlets

II.   **CONURBATIONS**
  (a)   Traditional industrial areas: Birmingham
  (b)   Service/port areas: Liverpool, Newcastle, *Birkenhead*
  (c)   Service/manufacturing: Manchester, Nottingham, Leeds, *Salford*

III.   **SERVICE CENTRES**
  (a)   Commercial: Bristol, Reading, Southampton, Cardiff, Portsmouth, *Plymouth, Brighton*
  (b)   Professional and administrative: Oxford, Bedford, *Cambridge, Windsor/Maidenhead, Colchester, Chelmsford, St. Albans, Guildford*

IV.   **INDUSTRIAL CENTRES**
  (a)   Textiles: Bradford, Bolton, Blackburn, Rochdale, Oldham, Halifax, Dewsbury, Ashton-under-Lyne, Huddersfield, *Burnley, Nelson, Bury* and services: Leicester, Preston
  (b)   Engineering
         and commerce: Northampton, Gloucester, *Peterborough, Ipswich*
         and railways: Coventry, Sheffield, Derby, *Swindon, Rochester, Doncaster, Wakefield*
  (c)   Heavy engineering: West Bromwich, Wolverhampton, Walsall, Dudley, Warley, *Rotherham, Stoke, Nuneaton*

V.   **SUBURBS AND SUBURBAN CENTRES**
  (a)   New towns: *Milton Keynes, Telford, Wokingham, Hemel Hempstead, Hitchin/Letchworth, Crawley*
  (b)   Working class industrial suburbs: *Stretford, Stockport*
  (c)   High status mixed residential: *Woking, Rugby, Leamington*
  (d)   Centres of rapid growth: *Aylesbury, Wycombe, Basingstoke, Solihull*

Note: Those settlements appearing in italics are new additions to the typology since 1971.

TABLE 2.8.   *The public housing stock of Blackburn in 1981*

Size	Type	Completed pre-1945 Number	Completed 1945–64 Number	Completed 1964 on Number
1-bedroom	House	0	0	0
	Flat	124	800	2041
	Bungalow	64	67	302
2-bedroom	House	914	627	840
	Flat	35	713	1133
	Bungalow	0	0	20
3-bedroom	House	1148	2223	2353
	Flat	0	93	350
	Bungalow	0	0	5
4-bedroom	All	13	117	205

Source: The Chartered Institute of Public Finance and Accountancy, 1982.

TABLE 2.9.   *Percentage of the population in each of the eight CES families*

Family	National	Blackburn	Family	National	Blackburn
1	10.7	0	5	29.5	24.4
2	12.7	10.2	6	4.8	0.9
3	11.5	3.4	7	5.8	9.7
4	20.1	52.2	8	4.9	0.9

Source: see text.

TABLE 2.10. *Percentage of the population in those thirty-five CES clusters represented in Blackburn*

	Cluster	Number of EDs	Population in those EDs	Percentage of the total population	Percentage of the national population
Family Two	11	3	1309	1.3	2.9
	12	11	5637	5.6	4.0
	13	3	1380	1.4	1.9
	14	2	1128	1.1	2.3
	15	2	1105	1.1	1.6
Family Three	17	1	530	0.5	1.7
	18	1	226	0.2	2.7
	19	4	1793	1.8	2.7
	20	1	188	0.2	2.3
	21	19	7352	7.3	2.0
	22	21	9399	9.4	1.3
	23	28	12236	12.3	1.8
Family Four	24	3	1159	1.2	2.7
	25	11	5089	5.1	2.4
	26	8	3675	3.7	2.0
	27	8	3996	4.0	1.7
	28	9	4949	4.9	3.2
	32	3	1751	1.8	2.4
	33	12	6166	6.2	2.5
	36	4	1666	1.7	0.6
	38	4	1904	1.9	1.8

TABLE 2.10.—continued

	Cluster	Number of EDs	Population in those EDs	Percentage of the total population	Percentage of the national population
Family Five	39	3	1646	1.6	1.6
	40	4	2404	2.4	1.6
	41	1	572	0.6	2.7
	42	4	2842	2.8	2.2
	43	3	1550	1.6	2.0
	44	2	1142	1.1	1.6
	45	4	2050	2.1	1.4
	46	2	1264	1.3	2.0
Family Six	49	1	860	0.9	1.9
	50	1	877	0.9	2.0
Family Seven	51	1	703	0.7	0.9
	52	19	10016	10.1	0.7
Family Eight	57	1	547	0.5	2.0
	58	1	574	0.6	1.1

Note: The characteristics of those clusters underlined are described in the text.

TABLE 2.11.   *The twenty largest Asian settlements in 1971 and 1981*

Rank	Asian population 1971[1]	Asian population 1981[2]
1	Greater London[5]	Greater London[6]
2	Birmingham	Birmingham
3	Leicester	Leicester
4	Bradford	Bradford
5	Wolverhampton	Wolverhampton
6	Coventry	Coventry
7	West Bromwich/Warley[4]	Huddersfield
8	Leeds	Leeds
9	Walsall	Slough
10	Manchester	Manchester
11	Slough[3]	Walsall
12	Bolton	Blackburn
13	Huddersfield	Bolton
14	Derby	Luton
15	Blackburn	Derby
16	Luton	West Bromwich/Warley
17	Nottingham	Nottingham
18	Rochdale	Rochdale
19	Sheffield	Oldham
20	Preston	Preston

Notes: [1] Indian, Pakistani, or East African born
[2] resident in a household, the head of which was born in India, Pakistan, Bangladesh, or East Africa
[3] Slough was not categorized as a separate entity in the 1971 county reports. This therefore represents an estimate
[4] Warley and West Bromwich were categorized as separate entities in 1971 but not in 1981
[5] Within the GLC the order of boroughs was Ealing, Brent, Hounslow, Newham, Barnet, and Croydon
[6] Within the GLC the order of boroughs was Ealing, Brent, Newham, Hounslow, Harrow, and Barnet

TABLE 2.12. *The percentage of the Asian population in ethnic settlements of different types, 1971 and 1981*

		1971 Asian	1971 NC	LQ	1981 Asian	1981 NC	LQ
I.	**LONDON**						
	(a) Traditional industrial areas	4.6	4.2	1.1	4.4	4.8	0.9
	(b) New industrial areas	2.4	1.0	2.4	1.6	1.4	1.1
	(c) Inner areas	19.9	30.5	0.6	16.5	22.9	0.7
	(d) Suburbs: mixed	10.1	6.8	1.5	11.7	11.6	1.0
	(d) Suburbs:exclusive	3.3	2.8	1.2	3.1	4.4	0.7
	(e) Service	8.6	13.9	0.6	10.3	13.3	0.8
II.	**CONURBATIONS**						
	(a) Traditional industrial	10.1	8.7	1.2	10.8	7.3	1.5
	(b) Service/port	0.9	1.0	0.9	0.8	1.0	0.8
	(c) Service/textile	4.9	5.2	0.9	4.7	4.3	1.1
III.	**SERVICE CENTRES**						
	(a) Commercial	1.9	3.0	0.6	1.6	3.3	0.5
	(b) Prof. and admin.	1.1	1.0	1.1	1.1	2.3	0.5
IV.	**INDUSTRIAL CENTRES**						
	(a) Textile	11.9	7.5	1.6	14.0	8.5	1.6
	Textile and services	5.9	3.6	1.6	6.9	4.1	1.7
	(b) Engineering and commerce	0.4	0.6	0.7	0.5	1.3	0.4
	Engineering and railways	5.2	3.9	1.3	5.1	4.0	1.3
	(c) Heavy engineering	8.6	6.3	1.4	6.7	5.4	1.2

TABLE 3.1.   *The distribution of the Asian population, 1961*

	% of total Indian population	% of total Pakistani population
N. Region	2.64	3.93
Tyneside conurbation	0.95	1.52
Remainder	1.68	2.41
East & West Ridings Region	4.56	19.73
W. Yorks. conurbation	2.67	16.29
Remainder	1.89	3.44
NW Region	6.43	6.98
SE Lancs conurbation	2.81	3.66
Merseyside conurbation	1.43	0.79
Remainder	2.19	2.53
N. Midlands Region	4.84	3.79
Midland Region	11.8	24.94
W. Midlands conurbation	7.46	22.32
Remainder	4.21	3.36
Eastern Region	6.80	4.43
SE Region	46.45	27.72
Greater London	37.05	24.25
Remainder	9.41	3.47
S. Region	8.01	4.49
SW. Region	6.86	2.52
Wales	1.58	1.44

Source: Census England and Wales, 1961. *Birthplace and Nationality Tables*, HMSO, 1964. 'Birth-places, Nationalities and Citizenship of Residents of England and Wales born outside the British Isles'. pp. 4–38.

TABLE 3.2.   *The shifting sex balance of Asian settlement in the UK 1962–78*

	Indian sex ratio male:female	Pakistani sex ratio male:female
July–Dec. 1962	1:0.56	1:0.36
1963	1:0.36	1:0.10
1964	1:0.78	1:0.49
1965	1:1.17	1:0.86
1966	1:1.28	1:1.85
1967	1:1.27	1:2.59
1968	1:1.63	1:2.51
1969	1:2.01	1:2.88
1970	1:1.93	1:2.45
1971	1:1.93	1:2.14
1972	1:2.01	1:2.42
1973	1:7.52	1:4.85
1974	1:2.48	1:2.77
1975	1:1.64	1:2.49
1976	1:1.33	1:2.94
1977	1:2.07	1:4.78
1978	1:1.19	1:4.51

Source: Table 4, 'Number of men, women and children among persons admitted as long term visitors, students, Ministry of Labour voucher holders, dependants, and other persons coming for settlement'. *Immigration Statistics 1973–1978, Control of Immigration Statistics 1962–1972.* HMSO, Cmnd nos. 2379, 2658, 2979, 3258, 3594, 4029, 4327, 4951, 5285, 5603, 6504, 6883, 7160, 7565.

TABLE 3.3.   *Men as a percentage of all those who had come to the UK up to a given date*

Up to	Pakistani/Bangladeshi	Indian
1956	92	79
1959	93	70
June 1962	93	73
Dec. 1964	90	69
1966	86	64
1968	78	61
1969	73	59
1971	68	57
1974	65	56

Source: D.J. Smith (1977)

TABLE 3.4.   *The distribution of the Asian population, 1971*

	% of total Indian population	% of total Pakistani population
N. Region	1.53	2.07
Tyneside conurbation	0.54	0.61
Remainder	0.98	1.45
Yorks. & Humberside	6.25	18.88
W. Yorks conurbation	4.95	15.67
Remainder	1.31	3.21
NW. Region	7.29	14.19
Merseyside conurbation	0.59	0.30
SE. Lancs. conurbation	3.87	8.88
Remainder	2.83	5.00
E. Midlands	7.12	3.46
Notts/Derby	2.11	2.14
Remainder	5.00	1.55
W. Midlands	19.00	20.04
W. Midlands conurbation	13.87	16.77
Remainder	5.15	3.28
E. Anglia	1.26	1.26
SE Region	49.87	34.59
Greater London	33.03	21.53
Outer Metropolitan	10.55	10.00
Outer SE	6.28	3.06
SW Region	3.74	1.57
Wales	1.06	1.20
Scotland	2.84	2.70
Central Clydeside conurbation	1.13	1.75
Remainder	1.70	0.95
Total	99.96	99.96

Source: Census Great Britain, 1971, *Age, Marital Condition and General Tables*, HMSO, 1974. 'Population by sex, marital condition, area of enumeration, country of birth and whether visitor to UK or not', pp. 26–127.

TABLE 3.5. *Estimates of change mid-1971 to mid-1976; ethnic population of UK*

	Mid-1971	Births	Deaths	Net migration	Mid-1976	Increase
Indian	307	62	7	28	390	27
Pakistani	171	43	3	35	246	44
East African Asian	68	9	1	84	160	135

Note: all figures in thousands
Source: OPCS Immigrant Statistics Unit, 1977.

TABLE 3.6. *The distribution of the Asian population, 1981*

	% of total Indian population	% of total Pakistani population[1]
N. Region	1.32	1.88
Tyneside conurbation	0.61	0.80
Remainder	0.71	1.09
Yorks and Humberside	5.67	17.15
W. Yorks conurbation	4.46	14.24
S. Yorks conurbation	0.58	2.49
Remainder	0.62	0.42
N.W. Region	7.22	14.22
Merseyside conurbation	0.55	0.31
Greater Manchester	3.85	8.83
Remainder	2.81	5.07
E. Midlands	8.71	3.64
W. Midlands	18.47	18.95
W. Midlands conurbation	16.10	16.77
Remainder	2.37	2.18
E. Anglia	1.26	1.29
S.E. Region	50.69	37.06
Greater London	35.50	24.38
Outer Metropolitan	9.91	9.66
Outer S.E.	5.28	3.02
S.W. Region	3.29	1.50
Wales	1.04	1.27
Scotland	2.32	3.00
Central Clydeside conurbation	1.03	1.94
Remainder	1.29	1.06
Total:	99.99	99.96

Source: Census Great Britain, 1981. *Country of Birth*, HMSO, 1983
Note: [1] Pakistani population taken to include Bangladeshis.
All data refer to birthplace groups not ethnic groups.

TABLE 3.7.   *Age structure of the Asian population, 1971*

Male		India	Pakistan	East Africa
		%	%	%
Age	0–14	20	20	19
	15–24	9	14	19
	25–44	19	27	15
	45 +	6	7	1
	All	54	68	55
Female				
Age	0–14	18	15	18
	15–24	8	6	17
	25–44	14	9	10
	45 +	5	2	1
	All	45	32	45
Total		100	100	100
Number in '000s		303	169	67

Source: OPCS Immigrant Statistics Unit, 1977.

TABLE 3.8.   *Take-up of offers of housing to URB by region*

	Offers (1/1/73)	Occupied	% take-up
North	220	65	29.5
Yorkshire	218	132	60.6
E. Midlands	171	91	53.2
Scotland	301	150	49.8
Wales	95	61	64.2
South West	214	139	64.9
South East	357	233	81.8
East Anglia		59	
West Midlands	196	124	63.3
North West	249	159	63.8
London	75	58	77.3
Total	2096	1271	

Source: Swinerton *et al.* (1975)

TABLE 3.9. *Regional distribution of those Ugandans resettled by the URB*

Region	Families	
	Private Accom.	Local Authority Accom.
North	9	86
Yorkshire	54	173
E. Midlands	51	121
Scotland	13	190
Wales	12	75
South West	129	171
South East	229	332
East Anglia	24	77
W. Midlands	60	173
North West	56	215
Greater London	214	180
Total	851	1793

Source: *Uganda Resettlement Board Final Report (1974)*, Appendix C, p. 24

TABLE 3.10   *The distribution of the East African Asian population, 1981*

	% of East African birthplace group[1]
N. Region	0.94
Tyneside conurbation	0.32
Remainder	0.62
Yorks and Humberside	3.30
W. Yorks. conurbation	2.29
S. Yorks. conurbation	0.39
Remainder	0.62
NW. Region	5.99
Merseyside conurbation	0.39
Greater Manchester conurbation	3.46
Remainder	2.14
E. Midlands	13.18
W. Midlands	8.67
W. Midlands conurbation	7.04
Remainder	1.63
E. Anglia	1.69
SE. Region	59.94
Greater London	46.45
Outer Metropolitan	8.93
Outer SE.	4.56
SW. Region	2.98
Wales	1.06
Scotland	2.23
Central Clydeside conurbation	0.52
Remainder	1.72
TOTAL:	99.98

Source: Census Great Britain, 1981.
Note: [1] this birthplace group contains individuals who are not of Asian ethnicity.

TABLE 7.1. *Stated reasons for migration to Britain*

	East African Asians	South Asians
Political problems	31	0
To get a job/career advancement	26	135
Marriage	5	10
Education	11	20
To visit	3	1
Relatives/Friends here	14	61
Sent by parents	1	0
To improve life	13	28
Financial reasons	0	22
Direct recruitment	0	3
Other people coming	0	3
For a change	0	1
Don't know	0	4
Total	104	288

Source: 1978 Sample Survey

TABLE 7.2.   *Date of migration to the UK*

	East African Asians	South Asians
1951	0	1
1953	0	0
1954	0	1
1955	0	0
1956	0	2
1957	0	7
1958	3	4
1959	0	5
1960	0	17
1961	0	32
1962	0	27
1963	1	22
1964	4	16
1965	5	13
1966	3	28
1967	11	19
1968	13	25
1969	5	14
1970	6	6
1971	3	4
1972	12	2
1973	4	1
1974	6	0
1975	8	1
1976	14	6
1977	2	2
1978	0	2
Total	100	257

Source: 1978 Sample Survey
Note: for significance of time periods indicated by lines, see text

TABLE 7.3.   *The age structure of the adult South and East African Asian populations in Blackburn*

Age	East African Asian	South Asians
16–20	216	510
21–25	237	669
26–30	177	441
31–35	52	324
36–40	80	395
41–45	67	296
46–50	62	203
51–55	42	82
56–60	24	39
61–65	28	26
66–70	28	4
71 +	10	8
Total	1023	2997

Source: 1977 Asian Census. Only contains those willing, or able, to declare their age

TABLE 7.4.   *The occupational structure of the South and East African Asian populations of Blackburn prior to migration to Britain*

Occupation	East African Asians	South Asians
Professions	1	8
Management	4	0
Priest	1	3
Business	38	18
Public Service	0	3
Clerk	5	2
Printing	1	3
Clothing	0	4
Engineering	2	3
Radio technician	1	1
Motor repairs	2	0
Other skilled	3	6
Construction	2	0
Textiles	0	3
Semi-skilled		
Factory work	1	4
Messenger	0	1
Farmer/labourer	6	81
Forces	0	8
Transport	3	6
Students	27	83
Total	96	237

Source: 1978 Sample Survey

TABLE 8.1. *The percentage of the total Asian population in EDs of varying concentration: Blackburn, 1971, 1977*

Location Quotients	% Asians found in such EDs 1971[a]	1977[b]
0–0.99	7.3	9.5
1.0–1.99	6.1	14.0
2.0–2.99	7.8	10.5
3.0–3.99	16.3	11.5
4.0–4.99	9.8	2.8
5.0 +	52.8	51.6
	n = 4884	n = 9461

Sources: [a] 1971 Small Area Statistics. OPCS data from the national census (unpublished)
[b] Asian Census plus additions for non-response

TABLE 8.2. *Ward level IDs between the Asian and white populations of Blackburn: 1968–84*

Year	ID	Year	ID
1968	50.37	1976	56.14
1969	51.98	1977	55.39
1970	54.27	1978	56.92
1971	56.20	1979	54.98
1972	58.71	1980	55.67
1973	58.74	1981	53.58
1974	56.96	1982	53.34
1975	55.96	1983	52.87
		1984	51.80

Source: Calculated from counts of Asian names on electoral register

TABLE 8.3. *Ward level $a^{P*}a$ between Asians, Blackburn: 1968–84*

Year	$a^{P*}a$	Expected $a^{P*}a$
1968	0.05	0.02
1970	0.11	0.03
1972	0.14	0.04
1974	0.18	0.06
1976	0.21	0.07
1978	0.26	0.09
1980	0.31	0.10
1982	0.32	0.10
1984	0.32	0.11

Source: Calculated from counts of Asian names on Electoral register

TABLE 8.4. *Comparisons of the tenure breakdown of Blackburn's Asian population with national data*

	Asians in Blackburn 1977[a] %	South Asians in Britain 1978[b] %	Asians in Britain 1975–8[c] %
Owner occupation	94.3	69.9	72
Private renting, furnished	3.1	12.7	10
Private renting, unfurnished	1.2	6.9	6
Council renting	0.5	10.1	8
Others	0.8	0.4	

Source: [a] 1977 Asian Census
          [b] National Dwelling and Housing Survey, 1978, Table 8, p. 33
          [c] General Household Survey, 1978, p. 26

TABLE 8.5. *Industrial structure of the Asian population in Blackburn, 1977*

Industrial category	Number employed	% of economically active
Textiles	1231	55.0
Electrical engineering	253	11.3
Clothing	139	6.2
Distribution	118	5.3
Paper and printing	77	3.4
Metal goods	76	3.4
Transport	66	2.9
Mechanical engineering	52	2.3
Food and drink	48	2.1
Professional/Scientific	42	1.9
Metal manufacture	34	1.5
Miscellaneous services	30	1.3
Chemical and allied	27	1.2
Finance/Insurance/Banking	16	0.7
Leather goods	9	0.4
Public administration	6	0.3
Vehicles	4	0.2
Iron and Steel	4	0.2
Construction	3	0.1
Agriculture	1	
Timber/Furniture	1	
Other manufacturing	1	
Total	2238	99.7

Source: 1977 Asian Census

TABLE 9.1.   *Remittances by overseas Pakistanis*

Year	Amount (£m)	Year	Amount (£m)
1970–1	24.35	1977–8	578.15
1971–2	53.45	1978–9	689.5
1972–3	61.9	1979–80	871.5
1973–4	69.25	1980–1	1048.6
1974–5	106.1	1981–2	1112.0
1975–6	167.1	1982–3	1443.1
1976–7	288.85	1983–4	1368.7

Source: Unpublished data supplied by Government of Pakistan.

TABLE 9.2.   *Remittances by Pakistanis in Britain*

Year	Amount (£m)
1973	19.77
1974	19.32
1975	24.77
1976	19.98
1977	na
1978	29.18
1979	44.56
1980	61.63
1981	48.52
1982	86.20
1983	67.76

Source: Unpublished data supplied by Government of Pakistan.

TABLE 9.3.   *Official recorded remittances by overseas Indians*

Year	£m	Year	£m
1960–1	2.64	1970–1	8.02
1961–2	2.43	1971–2	10.26
1962–3	2.36	1972–3	9.72
1963–4	2.79	1973–4	11.95
1964–5	3.30	1974–5	16.46
1965–6	5.58	1975–6	31.80
1966–7	6.52	1976–7	43.80
1967–8	7.10	1977–8	60.50
1968–9	8.48	1978–9	62.30
1969–70	8.19		

Note: 1978–9 represented the most recent statistics available in March 1982.
Source: Unpublished data supplied by Reserve Bank of India.

TABLE 9.4.   *Enumeration district level indices of dissimilarity for households; East African and South Asian ethnic groups, Blackburn C.B. 1977*

Indian-East African	26.3	Pakistani-East African	52.3
Indian-Pakistani	52.6	Pakistani-Indian	52.6
Indian-White	76.9	Pakistani-White	84.9
n = 809		n = 504	
East African-Indian	26.3	White-Indian	76.9
East African-Pakistani	52.3	White-East African	82.3
East African-White	82.3	White-Pakistani	84.9
n = 367		n = 92074 individuals.	

Source: 1977 Asian Census.

TABLE 9.5.   *Ward level indices of dissimilarity for adult individuals East African and South Asian birthplace groups, Blackburn C.B., 1977*

Indian-East African	14.2	Pakistani-Indian	34.5
Indian-Pakistani	34.5	Pakistani-East African	39.7
Indian-White	53.6	Pakistani-White	62.0
n = 2479		n = 1450	
East African-Indian	14.2	White-East African	37.6
East African-White	37.6	White-Indian	53.6
East African-Pakistani	39.7	White-Pakistani	62.0
n = 743		n = 92074	

Source: Robinson, 1979c.

TABLE 9.6.   *Ward level P* indices: Blackburn C.B., 1977*

	P*	Expected P*	Percentage
Indian	0.25	0.09	277
Pakistani	0.13	0.03	433
East African	0.08	0.02	400

Note: for adult individuals and their children.
Source: 1977 Asian Census.

TABLE 9.7.   *The employment structure of the South and East African Asian minorities in Blackburn, 1977*

Industrial category	South Asian Persons[a]	Percentage	East African Asian Persons[a]	Percentage
Agriculture	1	0.07		
Food, drink and tobacco	24	1.67	11	2.03
Chemicals	25	1.73	8	1.48
Metal manufacture	10	0.69	3	0.55
Mechanical engineering	84	5.83	57	10.51
Electrical engineering	158	10.96	88	16.24
Vehicles	1	0.07	3	0.55
Textiles	814	56.49	203	37.45
Leather and Furs	6	0.42		
Clothing and Footwear	53	3.68	70	12.91
Glass	1	0.07		
Timber and Furniture	5	0.35		
Paper and Printing	65	4.51	16	2.95
Construction	5	0.35	5	0.92
Transport	53	3.68	17	3.14
Distributive trades	62[b]	4.30	37[b]	6.82
Insurance, banking and finance	12	0.83	5	0.92
Professional and scientific	39	2.71	12	2.21
Miscellaneous services	11	0.76	2	0.37
Public administration	11	0.76	5	0.92

Notes: [a] includes males and females
         [b] South Asians enumerated all family members as shopkeepers; East Africans only head of household.
Source: 1977 Asian Census

TABLE 9.8.   *The Social Class structure of the Asian populations of Blackburn, 1977*

Social Class	Percentage East African Asians	Percentage South Asians
1	0.6	1.0
2	4.5	2.9
3	37.1	33.2
4	47.1	49.3
5	10.6	13.6
Total	99.9	100
	n = 310	n = 1188

Note: Heads of household only
Source: 1977 Asian Census.

TABLE 10.1.   *The composition of Blackburn's Asian minority*

Group	Households	Percentage
Rural Indian Gujurati Muslims	505	29.7
Rural Pakistani Punjabi Muslims	361	21.2
Urban Indian Gujurati Muslims	149	8.7
Rural East African Gujurati Muslims	140	8.2
Urban East African Gujurati Muslims	95	5.6
Urban Pakistani Punjabi Muslims	94	5.5
East African Gujurati Hindus	74	4.3
Indian Urdu-speaking Muslims	47	2.8
Punjabi speakers not specified	40	2.3
Indian Gujurati Hindus	40	2.3
Pakistani Urdu-speaking Muslims	34	1.9
East African Urdu-speaking Muslims	24	1.4
All Marathi speakers	23	1.4
Punjabi Sikhs	23	1.4
All Bengalis	18	1.1
Other Hindus	15	0.9
Others	19	1.1
Total	1702	99.8

Source: 1977 Asian Census

TABLE 10.2.   *Structural integration; inter-ethnic residential segregation* [a]

Groups [b]	1	2	3	4	5	6	7	8	9	10	11	12
1	—	23.5	26.4	37.9	34.6	24.5	57.5	81.8	53.4	53.6	60.0	59.7
2		—	27.3	20.4	33.9	34.8	56.2	82.7	44.8	44.9	51.7	57.7
3			—	34.4	54.5	36.9	53.1	77.6	44.7	34.3	54.5	57.9
4				—	41.3	43.5	61.1	85.0	59.5	38.9	60.5	71.9
5					—	32.8	62.6	77.4	48.9	63.3	68.0	53.3
6						—	60.0	75.6	42.1	54.6	62.8	50.9
7							—	33.8	49.9	57.5	54.3	36.8
8								—	78.9	79.9	81.9	56.3
9									—	30.4	26.9	28.1
10										—	33.5	45.9
11											—	45.1
12												—

Notes: [a] Indices of Dissimilarity at Ward level for households by ethnicity

[b] 1 = Rural East African Gujurati Muslims
2 = Urban East African Gujurati Muslims
3 = Rural Indian Gujurati Muslims
4 = Urban Indian Gujurati Muslims
5 = East African Urdu-speaking Muslims
6 = Indian Urdu-speaking Muslims
7 = East African Gujurati Hindus
8 = Indian Gujurati Hindus
9 = Pakistani Urdu-speaking Muslims
10 = Urban Pakistani Punjabi Muslims
11 = Rural Pakistani Punjabi Muslims
12 = Punjabi Sikhs

Source: 1977 Asian Census

TABLE 10.3.   *Relative ethnic association*

Groups[a]	Language	Secondary contact 1	Secondary contact 2	Associations	Primary contact	Orientation 1	Orientation 2	Orientation 3	Acculturation 1	Acculturation 2	Social class	Identification 1	Identification 2	Exogamy
EGH	1	5	9	2.5	6	7	11	4	5	2	10	3	5	9.5
EUI	3	2	5	2.5	4	1.5	1	1.5	1	2	3	12	2	1
IGH	12	8	11	2.5	3	9	7	11.5	9	8	12	1	2	9.5
IUI	9	11	1	7	10	8	4	1.5	3	7	11	11	9	9.5
PS	6	1	2	2.5	1	1.5	12	11.5	2	2	8	2	2	9.5
PUI	11	6	6	12	7.5	5	10	9	10	5	9	8	6	9.5
REGI	5	4	3	10	5	4	3	10	4	4	4	4	7	9.5
RIGI	8	10	4	9	11	11	6	7	8	6	6	9	10	5
RPPI	10	9	10	5	7.5	12	8	5	12	12	7	7	4	4
UEGI	2	3	7	11	2	3	2	3	6	11	2	5	8	3
UIGI	7	12	12	8	12	6	9	8	11	9	5	6	11	6
UPPI	4	7	8	6	9	10	5	6	7	10	1	10	12	2

Notes: Secondary contact 1 = industrial concentration
Secondary contact 2 = residential concentration
Orientation 1 = remittances
Orientation 2 = return migration
Orientation 3 = transmission of mother tongue to children
Acculturation 1 = use of traditional dress; adult women
Acculturation 2 = use of traditional dress; children
Identification 1 = social distance measure
Identification 2 = perceived friendliness of British
All scores represent ranks. Lower ranks indicate more assimilable groups
[a] See table 10.2 for full titles

TABLE 10.4.   *Biological amalgamation; inter-ethnic exogamy*[a]

Groups[b]	1	2	3	4	5	6	7	8	9	10	11	12
1	—	6.9	22.1	4.3	3.1	0	0	0	0	0	0.7	0
2		—	16.6	12.3	0	0	0	0	0	0	0	0
3			—	0	0	1.4	0	0	0	0	0.2	0
4				—	0	0	0	0	0	0	0	0
5					—	38.9	0	0	6.9	0	0	0
6						—	0	0	10.2	0	0.8	0
7							—	39.1	0	0	0	0
8								—	0	0	0	0
9									—	4.2	3.2	0
10										—	0	0
11											—	0
12												—

Notes: [a] Savorgnan's index for groups defined by birthplace, language and religion
           [b] 1 = Rural East African Gujurati Muslims
                2 = Urban East African Gujurati Muslims
                3 = Rural Indian Gujurati Muslims
                4 = Urban Indian Gujurati Muslims
                5 = East African Urdu-speaking Muslims
                6 = Indian Urdu-speaking Muslims
                7 = East African Gujurati Hindus
                8 = Indian Gujurati Hindus
                9 = Pakistani Urdu-speaking Muslims
               10 = Urban Pakistani Punjabi Muslims
               11 = Rural Pakistani Punjabi Muslims
               12 = Punjabi Sikhs
Source: 1977 Asian Census

TABLE 10.5.　Absolute ethnic association

Groups	Language	Secondary contact		Associations	Primary contact	Orientation			Acculturation		Social class	Identification		Exogamy
		1	2			1	2	3	1	2		1	2	
EGH	77	84	68	100	23	54	54	85	78	0	17	1.90	7.7	0
EUI	75	89	58	100	20	20	0	80	25	0	8	4.00	0	5.00
IGH	57	77	73	100	17	67	33	100	83	20	28	1.33	0	0
IUI	63	63	53	78	27	64	30	80	64	18	19	2.75	11.1	0
PS	69	92	54	100	0	20	60	100	50	0	13	1.50	0	0
PUI	58	83	58	60	25	50	42	92	91	14	15	2.60	8.3	0
REGI	71	84	54	71	22	45	29	94	73	8	9	2.23	9.4	0
RIGI	63	64	54	72	40	72	32	90	82	17	12	2.67	11.7	0.78
RPPI	59	65	69	82	25	72	35	86	99	46	13	2.49	6.4	0.86
UEGI	75	87	63	64	10	26	12	84	80	27	6	2.24	10	0.89
UIGI	68	55	74	73	42	52	36	90	93	22	11	2.29	13	0.57
UPPI	72	78	66	79	26	68	32	89	81	23	5	2.71	26	4.95

Notes: Categories as in table 10.3.
All scores are percentages, except Identification 1 which is an averaged social distance score. Minimum value 1 (absence of social distance), maximum 6 (complete social distance).
See text for interpretation of results.

BIBLIOGRAPHY

Abler, R.F., Adams, J.S., and Gould, P.R. (1971), *Spatial Organization: The Geographer's View of the World,* Englewood Cliffs: Prentice Hall.

Abrams, M. (1969), 'Attitudes of the British public', in *Colour and Citizenship* (ed. E.J.B. Rose), London: Oxford University Press.

Adorno, T.W. *et al.* (1964), *The Authoritarian Personality,* New York: John Wiley.

Airey, C. (1984), 'Social and moral values', in *British Social Attitudes: The 1984 Report* (eds. R. Jowell and C. Airey), Aldershot: Gower.

Aldrich, H. (1980), 'Asian shopkeepers as a middleman minority,' in *The Inner City: Employment and Industry* (eds. A.Evans and D. Eversley), London: Heinemann.

Aldrich, H., Cater, J., Jones, T.P., and McEvoy, D. (1981), 'Business development and self-segregation: Asian enterprise in three British cities'. in *Ethnic Segregation in Cities* (eds. C. Peach, V. Robinson, and S.J. Smith), London: Croom Helm.

Allport, G.W. (1954), *The Nature of Prejudice,* Cambridge, Mass.: Addison-Wesley.

Andrews, H.F. (1971), 'A cluster analysis of British towns,' *Urban Studies,* 8, 271–84.

Antonovsky, A. (1956), 'Toward a refinement of the "marginal man" concept', *Social Forces,* 35, 57–62.

Anwar, M. (1979), *The Myth of Return: Pakistanis in Britain,* London: Heinemann.

Armen, G. (1972), 'A classification of cities and city regions in England and Wales, 1966', *Regional Studies,* 6, 149–82.

Aurora, G.S. (1967), *The New Frontiersmen,* Bombay: Popular Prakastan.

Avison, E. (1965), 'Immigrants in a small borough', *Institute of Race Relations Newsletter,* Oct, 11, 5.

Bagley, C. (1970), *Social Structure and Prejudice in Five English Boroughs,* London: Institute of Race Relations.

Bagley, C. (1972), 'Interracial marriage in Britain—some statistics', *New Community,* 1, 318–26.

Baker, A.M. (1982), 'Ethnic enterprise and modern capitalism: Asian small businesses', *New Community,* 9, 478–86.

Ballard, C. (1978), 'Arranged marriages in the British context', *New Community,* 6, 181–97.

Ballard, R. and Ballard, C. (1977), 'The Sikhs—the development of South Asian settlement in Britain', in *Between Two Cultures* (ed. J.L. Watson), Oxford: Blackwell.

Barot, R. (1972), 'A Swaminarayan sect as a community', *New Community,* 2, 34–7.

Barth, F. (1969), *Ethnic Groups and Boundaries: The Social Organisation of Culture Difference,* London: Allen and Unwin.

Batchelor, C. (1984), 'A long way in a short time', *Financial Times,* 3 September.

Bater, J.H. (1980), *The Soviet City,* London: Edward Arnold.

Bell, C. (1977), 'On housing classes', *Australian and New Zealand Journal of Sociology,* 13, 36–40.

Bell, D. (1975), 'Ethnicity and social change', in *Ethnicity: Theory and Experience* (eds. N. Glazer and D.P. Moynihan), Cambridge, Mass.: Harvard University Press.

Bell, W. (1954), 'A probability model for the measurement of ecological segregation', *Social Forces,* 32, 357–64.

Bentley, S. (1972), 'Intergroup relations in local politics: Pakistanis and Bangladeshis', *New Community*, 2, 44–7.

Beshers, J.M. (1962), *Urban Social Structure,* New York: Free Press.

Bharati, A. (1970), 'A social survey', in *Portrait of a Minority: Asians in East Africa* (eds.D.P. Ghai and Y.P. Ghai), Nairobi: Oxford University Press.

Blackburn Borough (1975), *Housing Renewal Strategy.*

Blackburn Borogh (1979), *Housing Renewal Strategy Review.*

Blauner, R. (1969), 'Internal colonialism and ghetto revolt', *Social Problems*, 16, 393–408.

Boal, F.W. (1978), 'Ethnic residential segregation', in *Social Areas in Cities: Processes, Patterns and Problems,* (eds. D.T. Herbert and R.J. Johnston), Chichester: Wiley.

Bogardus, E.S. (1930), 'A race relations cycle', *American Journal of Sociology*, 36, 612–17.

Bombay Chamber of Commerce (1978), *Inward Remittances Gujurat—A Survey,* Bombay: Commerce Research Bureau.

Brah, A. (1978), 'South Asian teenagers in Southall; their perceptions of marriage, family and ethnic identity', *New Community*, 6, 197–206.

Braithwaite, L. (1960), 'The concept of the marginal man', *Annals of the New York Academy of Sciences,* 83, 816–36.

Breton, R. (1964), 'Institutional completeness of ethnic communities and the personal relations of immigrants', *American Journal of Sociology*, 70, 193–205.

Briggs, A. (1968), *Victorian Cities*, Harmondsworth: Penguin.

Bristow, M. (1976), 'Britain's response to the Ugandan Asian crisis', *New Community*, 5, 265–79.

Bristow, M. (1979), 'Ugandan Asians: racial disadvantage and housing markets in Manchester and Birmingham', *New Community*, 7, 203–16.

Bristow, M. and Adams, B.N. (1977), 'Ugandan Asians and the housing market in Britain', *New Community*, 6, 65–77.

Brooks, D. (1969), 'Who will go back?' *Race Today*, September, 132–4.

Brooks, D. and Singh, K. (1979a), 'Ethnic commitment versus structural reality', *New Community*, 7, 19–31.

Brooks, D. and Singh, K. (1979b), 'Pivots and presents; Asian brokers in British foundries', in *Ethnicity at Work* (ed. S. Wallman), London: Macmillan.

Brown, C. (1984), *Black and White Britain: The Third PSI Survey*, London: Heinemann.

Burke, G. (1961), *Towns in the Making*, London: Edward Arnold.

Burney, E. (1967), *Housing on Trial: A Study of Immigrants and Local Government,* London: Oxford University Press.

Cable, V. (1969), *Whither Kenyan Emigrants*, London: Fabian Society.

Calef, W. and Nelson, H.J. (1956), 'Distribution of Negro population in the United States', *Geographical Review*, 46, 82–97.

Cashmore, E. (1979), *Rastaman*, London: Allen and Unwin.

Cashmore, E. (1981), 'After the Rastas', *New Community,* 9, 173–82.

Cater, J. (1981), 'The impact of Asian estate agents on patterns of ethnic residence: a case study in Bradford', in *Social Interaction and Ethnic Segregation* (eds. P. Jackson and S.J. Smith), London: Academic Press.

Cater, J. and Jones, T.P. (1978), 'Asians in Bradford', *New Society*, 13 April, 81–2.

Cater, J.C., Jones, T.P., and McEvoy, D. (1979), 'Minority business as a social ladder: a case study', paper submitted to *Sociological Review*.

Chansarkar, B.A. (1973), 'A note on the Maratha community in Britain', *New Community*, 2, 302–6.

Child, I. (1943), *Italian or American?* New Haven: Yale University Press.

Clemente, F. and Sturgis, R.B. (1971), 'Population size and industrial diversification', *Urban Studies*, 8, 65–8.

Collard, D. (1972), *Price, Prejudice and Exclusion in the Housing Market*, Bristol: SSRC Research Unit on Ethnic Relations.

Collard, D. (1973), 'Price and prejudice in the housing market', *Economic Journal*, 83, 510–15.

Commission for Racial Equality (1981), *Racial Harassment on Local Authority Housing Estates*, London: CRE.

Commission for Racial Equality (1984), *Race and Council Housing in Hackney*, London: CRE.

Commission on Industrial Relations (1970), *Report No. 4 on Birmid Qualcast*, London: HMSO.

Community Relations Commission (1976), *Between Two Cultures*, London: CRC.

Community Relations Commission (1977), *Housing Choice and Ethnic Concentration*, London: CRC.

Cox, O.C. (1948), *Caste, Class and Race*, New York: Doubleday.

Cullingworth Committee (1969), *Council Housing. Purposes, Procedures and Priorities*, London: HMSO.

Dahrendorf, R. (1959), *Class and Class Conflict in Industrial Society*, London: Routledge & Kegan Paul.

Dahya, B. (1972), 'Pakistanis in England', *New Community*, 2, 25–34.

Dahya, B. (1973), 'Pakistanis in Britain: transients or settlers?' *Race*, 14, 241–77.

Dahya, B. (1974), 'The nature of Pakistani ethnicity in industrial cities in Britain', in *Urban Ethnicity* (ed. A. Cohen), London: Tavistock.

Daniel, W.W. (1968), *Racial Discrimination in England*, Harmondsworth: Penguin.

Davey, A.G. and Norburn, M.V. (1980), 'Ethnic awareness and ethnic differentiation amongst primary school children', *New Community*, 8, 51–61.

Davison, B. (1968), 'No place back home: a study of Jamaicans returning to Kingston, Jamaica', *Race*, 9, 499–509.

Davison, R.B. (1966), *Black British*, London: Oxford University Press.

Delf, G. (1963), *Asians in East Africa*, London: Oxford Univesity Press.

Department of Employment (1979), *Employment Gazette*, London: HMSO.

Department of the Environment (1978), *National Dwelling and Housing Survey, 1978*, London: HMSO.

Desai, R. (1963), *Indian Immigrants in Britain*, London: Oxford University Press.

Dhanjal, B. (1976), 'Sikh women in Southall', *New Community*, 5, 109–15.

Dickie-Clarke, H.F. (1966), *The Marginal Situation: A Sociological Study of a Coloured Group*, London: Routledge & Kegan Paul.

Dines, M. (1973), 'Ugandan Asians one year on: cool reception', *New Community*, 12, 380–3.

Doherty, J. (1969), 'The distribution and concentration of immigrants in London', *Race Today*, 227–32.

Donnison, D. and Soto, P. (1980), *The Good City: A Study of Urban Development and Policy in Britain*, London: Heinemann.

Duncan, O.D. and Duncan, B. (1955), 'A methodological analysis of segregation indexes', *American Sociological Review*, 20, 210–17.

Duncan, O.D. and Lieberson, S. (1959), 'Ethnic segregation and assimilation', *American Journal of Sociology*, 64, 364–74.

Duncan, S.S. (1974), 'Cosmetic planning or social engineering? Improvement grants and improvement areas in Huddersfield', *Area*, 6, 259–69.

Duncan, S.S. (1977), *Housing Disadvantage and Residential Mobility: Immigrants and Institutions in a Northern Town,* Brighton: University of Sussex, Department of Urban and Regional Studies.

Economist (1981), 'Where will the jobs come from?' 3 January, 45–62.

Economist Intelligence Unit (1961), *Studies on Immigration from the Commonwealth, No. 1, Basic Statistics,* London: Economist.

Economist Intelligence Unit (1962), *Studies on Immigration from the Commonwealth, No. 2, The Immigrant Communities,* London: Economist.

Eyles, J. (1979), 'Area-based policies for the inner-city: context, problems and prospects', in *Social Problems and the City: Geographical Perspectives* (eds. D.T. Herbert and D.M. Smith), London: Oxford University Press.

Faux, R. (1980), 'From the Punjab to the Western Isles', *The Times,* 28 October.

Fenton, M. (1973), 'Costs of discrimination in the owner-occupied sector', *New Community,* 6, 279–83.

Fenton, M. (1977), *Asian Households in Owner Occupation: A Study of the Pattern, Costs, and Experiences of Households in Greater Manchester,* Bristol: SSRC Research Unit on Ethnic Relations.

Festinger, L., Schachter, S., and Back, K. (1950), *Social Pressures in Informal Groups,* New York: Harper.

Field, S., Mair, B., Rees, T., and Stevens, P. (1981), *Ethnic Minorities in Britain: A Study of Trends in their Position since 1961,* London: HMSO.

Flett, H. (1979), *Black Council Tenants in Birmingham,* Bristol: SSRC Research Unit on Ethnic Relations.

Forester, T. (1978), 'Asians in business', *New Society,* 23 February.

Forster, E.M. (1924), *A Passage to India,* London: Edward Arnold.

Freeman, J. (1977), 'Small Heath: lessons from an inner area study', *The Planner,* 63, 46–7.

Freeman, T.W., Rodgers, H.B., and Kinvig, R.H. (1966), *Lancashire, Cheshire and the Isle of Man,* London: Nelson and Sons.

French, R.A. and Hamilton, F.E.I. (1979), *The Socialist City. Spatial Structure and Urban Policy,* Chichester: John Wiley.

Frucht, R. (1968), 'Emigration, remittances and social change', *Anthropologica,* 10, 193–208.

Fryer, P. (1984), *Staying Power: The History of Black People in Britain,* London: Pluto Press.

Gans, H.J. (1962), *The Urban Villagers,* New York: Free Press.

Gans, H.J. (1979), 'Symbolic ethnicity', *Ethnic and Racial Studies,* 2, 1–20.

General Household Survey, 1978 (1980), London: HMSO.

Ghai, D.P. (1970), 'An economic survey', in *Portrait of a Minority* (eds. D.P. Ghai and Y.P. Ghai), Nairobi: Oxford University Press.

Ghai, D.P. and Ghai Y.P. (1970), *Portrait of a Minority. Asians in East Africa,* Nairobi: Oxford University Press.

Ghuman, P.A.S. (1980), 'Bhattra Sikhs in Cardiff: family and kinship organisation', *New Community,* 8, 308–16.

Gibbs, J. and Martin, W. (1962), 'Urbanisation, technology and the division of labour', *American Sociological Review,* 27, 667–77.

Gibson, M. and Langstaff, M. (1977), 'Scope for improvement? Policies and strategies in gradual renewal', *The Planner,* 63, 35–8.

Goldberg, M.M. (1941), 'A qualification of the marginal man theory', *American Sociological Review,* 6, 52–8.

Goldthorpe, J.H. (1980), *Social Mobility and Class Structure in Modern Britain,* London: Oxford University Press.

Golovensky, D.I. (1952), 'The marginal man concept: an analysis and critique', *Social Forces*, 30, 333–9.

Goodall, J. (1966), *Institute of Race Relations Newsletter Supplement on Huddersfield*, London: Institute of Race Relations.

Gordon, M.M. (1964), *Assimilation in American Life*, New York: Oxford University Press.

Gordon, M.M. (1975), 'Toward a general theory of racial and ethnic group relations', in *Ethnicity: Theory and Experience* (eds. N. Glazer and D.P. Moynihan), Cambridge, Mass.: Harvard University Press.

Greater London Council (1976), *Race and Council Housing*, London: GLC.

Green, A.W. (1947), 'A re-examination of the marginal man concept', *Social Forces*, 26, 167–71.

Griffith, J.A.G. (1960), *Coloured Immigrants in Britain*, London: Oxford University Press.

Hahlo, K.G. (1980), 'Profile of a Gujurati community in Bolton', *New Community*, 8, 295–307.

Hallam R. (1972), 'The Ismailis in Britain', *New Community*, 1, 383–88.

Harrison, P. (1984), 'How race affects council housing', *New Society*, 12 January, 43–5.

Hartshorne, R. (1938), 'Racial maps of the United States', *Geographical Review*, 28, 276–88.

Hatch, J.C.S. (1971), *Constraints on Immigrant Housing Choice: Estate Agents*, Bristol: SSRC Research Unit on Ethnic Relations.

Hatch, J.C.S. (1973), 'Estate Agents as Urban Gatekeepers', unpublished paper delivered at the British Sociological Association Conference at Stirling.

Hechter, M. (1975), *Internal Colonialism. The Celtic Fringe in British National Development*, London: Routledge & Kegan Paul.

Hechter, M. (1978), 'Group formation and the cultural division of labour', *American Journal of Sociology*, 84, 293–318.

— Helweg, A.W. (1979), *Sikhs in England. The Development of a Migrant Community*, Delhi: Oxford University Press.

— Helweg, A.W. (1983), 'Emigrant remittances: their nature and impact on a Punjabi village', *New Community*, 10, 435–44.

Hill, C.S. (1965), *How Colour Prejudiced is Britain?* London: Gollancz.

Howard, E. (1902), *Garden Cities of Tomorrow*, London: Faber.

Hughes, E.C. (1949), 'Social change and status protest: an essay on the marginal man', *Phylon*, 10, 58–65.

Husain, M.S. (1975), 'The increase and distribution of New Commonwealth immigrants in Greater Nottingham', *East Midland Geographer*, 6, 105–29.

Husbands, C.T. (1975), 'The National Front: a response to crisis?' *New Society*, 15 May, 44–6.

Iliffe, L. (1978), 'Estimated fertility rates of Asian and West Indian immigrant women in Britain', *Journal of Biosocial Science*, 10, 189–97.

Immigrant Statistics Unit of OPCS (1975), 'Country of birth and colour', *Population Trends*, 2, 2–8.

Immigrant Statistics Unit of OPCS (1977), 'New Commonwealth and Pakistani population estimates', *Population Trends*, 9, 4–7.

Immigrant Statistics Unit of OPCS (1978), 'Marriage and birth patterns among the New Commonwealth and Pakistani population', *Population Trends*, 11, 5–9.

Immigrant Statistics Unit of OPCS (1979), 'Population of New Commonwealth and Pakistani ethnic origin: new projections', *Population Trends*, 16, 22–8.

Israel, W.H. (1964), *Colour and Community - A Study of Coloured Immigrants and Race Relations in an Industrial Town*, Slough: Slough Council of Social Service.

Jeffery, P. (1972), 'Pakistani families in Bristol', *New Community*, 1, 364–70.

Jeffery. P. (1973), 'Pakistani Families and their Networks in Bristol and Pakistan', Ph. D. thesis, University of Bristol.

Jeffery, P. (1976), *Migrants and Refugees. Muslim and Christian Pakistani Families in Bristol*, Cambridge: Cambridge University Press.

Jenkins, S. (1971), *Here to Live: a Study of Race Relations in an English Town*, London: Runnymede Trust.

John, D.W. (1969), *Indian Workers' Associations in Great Britain*, London: Oxford University Press.

Jones, F.L. (1967), 'Ethnic concentration and assimilation: an Australian case study', *Social Forces*, 45, 412–23.

Jones, H.R. and Davenport, M. (1972), 'The Pakistani community in Dundee. A study of its growth and demographic structure', *Scottish Geographical Magazine*, 88, 74–85.

Jones, P. (1980), 'Housing action areas: the facts', *Housing and Planning Review,* Spring, 5–8.

Jones, P. (1981a), 'British unemployment cycles and immigration: two case studies', *New Community*, 9, 112–17.

Jones, P. (1981b), 'Ins and outs of Home Office and IPS migration data: a reply', *New Community*, 9, 283–4.

Jones, P.N. (1967), *The Segregation of Immigrant Communities in the City of Birmingham, 1961*, Hull: University of Hull.

Jones, P.N. (1970), 'Some aspects of the changing distribution of coloured immigrants in Birmingham, 1961–66', *Transactions, Institute of British Geographers*, 50, 199–219.

Jones, P.N. (1978), 'The distribution and diffusion of the coloured population in England and Wales, 1961–71', *Transactions, Institute of British Geographers*, 3, 515–33.

Jones, T.P. (1982), 'Small business development and the Asian community in Britain', *New Community*, 9, 467–78.

Jones, T.P. and McEvoy, D. (1974), 'Residential segregation of Asians in Huddersfield', unpublished paper presented at the Institute of British Geographers' Annual Conference, Norwich.

Jones, T.P. and McEvoy, D. (1978), 'Race and space in cloud-cuckoo-land', *Area*, 10, 162–6.

Kanitkar, H.A. (1972), 'An Indian Elite in Britain', *New Community*, 1, 378–83.

Kannan, C.T. (1978), *Cultural Adaptation of Asian Immigrants*, Greenford: Kannan Publishing.

Kantrowitz, N. (1981), 'Ethnic segregation: social reality and academic myth', in *Ethnic Segregation in Cities* (eds. C. Peach, V. Robinson, and S.J. Smith), London: Croom Helm.

Karn, V.A. (1978), 'The financing of owner-occupation and its impact on ethnic minorities', *New Community*, 6, 49–65.

Karn, V.A. (1981), 'Race and Council Housing Allocations', unpublished paper presented at the Policy Seminar on Race Relations, Nuffield College, Oxford, 27 November.

Kearsley, G.W. and Srivastava, S.R. (1974), 'The spatial evolution of Glasgow's Asian community', *Scottish Geographical Magazine*, 90, 110–24.

Kennedy, R.J.R. (1943), 'Premarital residential propinquity and ethnic endogamy', *American Journal of Sociology*, 48, 580–4.

Kerckhoff, A.C. and McCormick, T.C. (1955), 'Marginal status and marginal personality', *Social Forces*, 34, 48–55.

Khan, V.S. (1977), 'Mirpuri villagers at home and in Bradford', in *Between Two Cultures* (ed. J.L. Watson), Oxford: Blackwell.

King, R. (1978), 'Return migration: a neglected aspect of population geography', *Area*, 10, 175–82.

Kohler, D. (1973), 'Public opinion and the Ugandan Asians', *New Community*, 2, 194–7.

Koller, M.R. (1948), 'Residential propinquity of white mates at marriage in relation to age and occupation of males, Columbus Ohio, 1938 and 1946', *American Sociological Review*, 13, 613–16.

Kumar, S. (1973), 'Uganda Asians one year later: Manchester reception', *New Community*, 2, 386–8.

Kunz, E.F. (1973), 'The refugee in flight: kinetic models and forms of displacement', *International Migration Review*, 7, 125–46.

Kuper, J. (1975), 'The Goan community in Kampala', in *Expulsion of a Minority* (ed. M. Twaddle), London: Athlone Press.

Labour Force Survey, 1981 (1983), London: HMSO.

Lapiere, R.T. (1928), 'Race prejudice: France and Britain', *Social Forces*, 7, 102–11.

Lawless, P. (1977), 'Housing action areas: powerful attack or financial fiasco', *The Planner*, 63, 39–42.

Lawrence, D. (1974), *Black Migrants: White Natives. A Study of Race Relations in Nottingham*, Cambridge: Cambridge University Press.

Lee, T.R. (1973), 'Immigrants in London: trends in distribution and concentration, 1961–71' *New Community*, 2, 145–59.

Lee, T.R. (1977), *Race and Residence. The Concentration and Dispersal of Immigrants in London,* Oxford: Oxford University Press.

Le Lohé, M.J. (1976), 'The National Front and the General Elections of 1974', *New Community*, 5, 292–301.

Lemon, A. (1976), *Apartheid : A Geography of Separation*, Farnborough: Saxon House.

Lieberson, S. (1961), 'The impact of residential segregation on ethnic assimilation', *Social Forces*, 40, 52–7.

Lieberson, S. (1981), 'An asymmetrical approach to segregation', in *Ethnic Segregation in Cities* (eds. C. Peach, V. Robinson, and S.J. Smith), London: Croom Helm.

Little, K. (1948), *Negroes in Britain. A Study of Race Relations in English Society*, London: Routledge & Kegan Paul.

Lyman, S.M. and Scott, M.B., (1967), 'Territoriality: a neglected sociological dimension', *Social Problems*, 15, 236–49.

Lyon, M.H. (1972), 'Ethnicity in Britain: the Gujurati tradition', *New Community*, 2, 1–11.

Mangat, J.S. (1969), *A History of the Asian in East Africa*, London: Oxford University Press.

Mann, J.W. (1957), 'The Problems of the Marginal Personality', unpublished Ph. D. thesis, University of Natal.

Mann, P.H. (1965), *An Approach to Urban Sociology*, London: Routledge & Kegan Paul.

Manners, R. (1965), 'Remittances and the unit of analysis in anthropological research', *Southwestern Journal of Anthropology,* 21, 179–95.

Marsh, P. (1967), *The Anatomy of a Strike,* London: Institute of Race Relations.

Mayer, P. (1961), *Townsmen or Tribesmen*, Cape Town: Oxford University Press.

McCart, M. (1973), 'Ugandan Asians one year later: Wandsworth. Unsettled Ugandan refugees', *New Community*, 2, 383–6.

McCormick, B. (1981), *Housing Segregation and the Journey to Work for Blacks in the UK*, Southampton: University of Southampton.

Michaelson, M. (1979), 'The relevance of caste among East African Gujuratis in Britain', *New Community*, 7, 350–60.

Miller, N. and Brewer, M.B. (1984), *Groups in Contact: The Psychology of Desegregation*, London: Academic Press.

Mitchell, J.C. (1966), 'Theoretical orientations in African urban studies', in *The Social Anthropology of Complex Societies* (ed. M. Banton), London: Tavistock.

Mitchell, J.C. (1969), 'The concept and use of social networks', in *Social Networks in Urban Situations* (ed. J.C. Mitchell), Manchester: Manchester University Press.

Moore, R. (1969), 'Citizens or immigrants: the problem of racial integration in contemporary urban Britain', in *Interaction: Nine Studies of Human Groups* (ed. P. de Berker), Oxford: Cassirer.

Moore, R. and Wallace, T. (1975), *Slamming the Door*, London: Martin Robertson.

Moser, C.A. and Scott, W. (1961), *British Towns: A Statistical Study of Their Social and Economic Differences*, Edinburgh: Oliver and Boyd.

Mullins, D. (1979), 'Asian retailing in Croydon', *New Community*, 7, 403–5.

Mullins, D. (1980), 'Race and Retailing', unpublished paper presented at the Institute of British Geographers' Annual Conference, Lancaster.

Mumford, L. (1961), *The City in History*, Harmondsworth: Penguin.

Musil, J. (1968), 'The development of Prague's ecological structure,' in *Readings in Urban Sociology* (ed R.E. Pahl), London: Pergamon.

Nagra, J.S. (1982), 'Asian supplementary schools,' *New Community*, 9, 431–7.

New Community correspondent (1972), 'Asian minorities in East Africa and Britain,' *New Community*, 1, 406–20.

Northway, M.L. (1952), *A Primer of Sociometry*, Toronto: University of Toronto Press.

Nowikowski, S. and Ward, R. (1978), 'Middle class and British? An analysis of South Asians in suburbia', *New Community*, 7, 1–11.

Office of Population Censuses and Surveys (1984), *Monitor MN* 84/1, London: OPCS.

Park, R.E. (1928), 'Human migration and the marginal man', *American Journal of Sociology*, 33, 881–93.

Parker, J. and Dugmore, K. (1977), 'Race and allocation of public housing—a GLC survey', *New Community*, 6, 27–41.

Parkin, F. (1979), *Marxism and Class Theory: A Bourgeois Critique*, London: Tavistock.

Patterson, H.O. (1968), 'West Indian migrants returning home: some observations', *Race*, 10, 69–77.

Patterson, S. (1968), *Immigrants in Industry*, London: Oxford University Press.

Peach, C. (1965), 'West Indian migration to Britain: the economic factors', *Race*, 7, 31–47.

Peach, C. (1966), 'Factors affecting the distribution of West Indians in Great Britain', *Transactions, Institute of British Geographers*, 38, 151–63.

Peach, C. (1967), 'West Indians as a replacement population in England and Wales', *Social and Economic Studies*, 16, 289–94.

Peach, C. (1968), *West Indian Migration to Britain*, London: Oxford University Press.

Peach, C. (1975), *Urban Social Segregation*, London: Longman.

Peach, C. (1979), 'British unemployment cycles and West Indian immigration 1955–74', *New Community*, 7, 40–4.

Peach, C. (1981a), 'Ins and outs of Home Office and IPS migration data, *New Community*, 9, 117–20.

Peach, C. (1981b), 'Straining at gnats and swallowing camels,' *New Community*, 9, 284–6.

Peach, C. and Winchester, S.W.C. (1974), 'Birthplace, ethnicity and the enumeration of West Indians, Indians, and Pakistanis', *New Community*, 3, 386–94.

Peach, C. and Shah, S. (1980), 'The contribution of council house allocation to West Indian desegregation in London, 1961–71', *Urban Studies*, 17, 333–41.

Pearson, G. (1976), 'Paki-bashing in a north east Lancashire cotton town', in *Working Class Youth Culture* (eds. G. Mungham and G. Pearson), London: Routledge & Kegan Paul.

Petersen, W. (1970), 'A general typology of migration', in *Readings in the Sociology of Migration* (ed.C. Jansen), London: Oxford University Press.

Pettigrew, J. (1972), 'Some observations on the social system of the Sikh Jats', *New Community*, 1, 354–64.

Phillips, D. (1981), 'The social and spatial segregation of Asians in Leicester', in *Social Interaction and Ethnic Segregation* (eds. P. Jackson and S.J. Smith), London: Academic Press.

Philpott, S.B. (1968), 'Remittance obligations, social networks, and choice among Montserratian migrants in Britain', *Man*, 3, 465–76.

Philpott, S.B. (1977), 'The Montserratians: migration dependency and the mainte-nance of island ties in England', in *Between Two Cultures* (ed. J.L. Watson), Oxford: Blackwell.

Price, C. (1969), 'The study of assimilation', in *Migration* (ed. J.A. Jackson), Cambridge: Cambridge University Press.

Pryce, K. (1979), *Endless Pressure: A Study of West Indian Lifestyles in Britain*, Harmondsworth: Penguin.

Radcliffe (Lord) (1969), 'Immigration and settlement: some general considerations', *Race*, 11, 42–52.

Rapkin, C. and Grigsby, W.G. (1960), *The Demand for Housing in Racially Mixed Areas*, Berkeley: Report to the Commission on Race and Housing.

Rattansi, P.M. and Abdulla, M. (1970), 'An educational survey', in *Portrait of a Minority* (eds. D.P.Ghai and Y.P. Ghai), Nairobi: Oxford University Press.

Rex, J.A. (1961), *Key Problems of Sociological Theory*, London: Routledge & Kegan Paul.

Rex, J.A. (1968), 'The sociology of a zone of transition', in *Readings in Urban Sociology* (ed. R.E. Pahl), London: Pergamon.

Rex, J. A. (1970), *Race Relations in Sociological Theory,* London: Weidenfeld and Nicholson.

Rex, J.A. (1973), *Race, Colonialism and the City*, London: Routledge & Kegan Paul.

Rex, J.A. (1977), 'Sociological theory and the city,' *Australian and New Zealand Journal of Sociology*, 13, 218–23.

Rex, J.A. and Moore, R. (1967), *Race, Community and Conflict,* Oxford: Oxford University Press.

Rex, J.A. and Tomlinson, S. (1979), *Colonial Immigrants in a British City,* London: Routledge & Kegan Paul.

Richmond, A.H. (1968), 'Return migration of Britons from Canada', *Population Studies*, 22, 263–71.

Richmond, A.H. (1973), *Migration and Race Relations in an English City: A Study in Bristol*, London: Oxford University Press.

Roberts, T. and Gunby, D. (1973), 'Housing improvement policy', *Journal of the Royal Town Planning Institute*, 59, 84–5.

Robinson, V. (1979a), *The Segregation of Asians in a British City; Theory and Practice*, Oxford: Oxford University School of Geography.

Robinson, V. (1979b), 'Choice and constraint in Asian housing in Blackburn', *New Community*, 7, 390–7.

Robinson, V. (1980a), 'Correlates of Asian immigration to Britain', *New Community*, 8, 115–23.

Robinson, V. (1980b), 'Asians and council housing', *Urban Studies*, 17, 323–31.

Robinson, V. (1980c), 'The achievement of Asian children', *Educational Research*, 22, 148–50.

Robinson, V. (1980d), 'Patterns of South Asian ethnic exogamy in Britain', *Ethnic and Racial Studies*, 3, 427–43.

Robinson, V. (1980e), 'Lieberson's P* index: a case study evaluation', *Area* 12, 307–12.

Robinson, V. (1981a), 'The development of South Asian settlement in Britain and the Myth of Return,' in *Ethnic Segregration in Cities* (eds. C. Peach, V. Robinson, and S.J. Smith), London: Croom Helm.

Robinson, V. (1981b), *The Dynamics of Ethnic Succession: A British Case-study*, Oxford: Oxford University School of Geography.

Robinson, V. (1982a), 'An Analysis of the Spatial Structure of the Asian Community in Blackburn and its Social Implications', D. Phil. thesis, University of Oxford.

Robinson, V. (1982b), 'The assimilation of South and East African Asian immigrants in Britain,' in *The Demography of Immigrants and Minority Groups in the United Kingdom* (ed. D.A. Coleman), London: Academic Press.

Robinson, V. (1984a), 'Racial antipathy in South Wales and its social and demographic correlates', *New Community*, 12, 116–24.

Robinson, V. (1984b), 'Asians in Britain; a study in encapsulation and marginality', in *Geography and Ethnic Pluralism* (eds. C.G. Clarke, D. Ley, and C. Peach), London: George Allen and Unwin.

Robinson, V. (1985a), 'Temporal and spatial variation in support for extreme right wing parties in Britain, 1966–83', *Cambria*, 11, 1–20.

Robinson, V. (1985b), 'The Vietnamese reception and resettlement programme in the UK: rhetoric and reality', *Ethnic Groups*, forth coming.

Robinson, V. (1985c), 'The importance of remittances for South Asians at home and in the UK', in *Proceedings of the Sixth International Symposium on Asian Studies*, Hong Kong: Asian Research Services.

Robinson, V. (1986), 'Bridging the Gulf: the economic significance of South Asian labour migration to and from the Middle East', in *Return Migration and Regional Economic Problems* (ed. R.J. King), London: Croom Helm.

Robinson, V. and Flintoff, I. (1982), 'Asian retailing in Coventry', *New Community*, 10, 251–9.

Robinson, W.S. (1950), 'Ecological correlations and the behaviour of individuals', *American Sociological Review*, 15, 351–7.

Rodgers, H.B. (1962), 'The changing geography of the Lancashire cotton industry', *Economic Geography*, 38, 299–314.

Rogerson, C.M. (1980), 'Internal colonialism, transnationalisation, and spatial inequality', *South African Geographical Journal*, 62, 103–20.

Rose, E.J.B. (1969), *Colour and Citizenship,* London: Oxford University Press.

Runnymede Trust (1975), *Race and Council Housing in London*, London: Runnymede Trust.

Runnymede Trust and Radical Statistics Race Group (1980), *Britain's Black Population*, London: Heinemann.

Savorgnan, F. (1950), 'Matrimonial selection and the amalgamation of heterogeneous groups', *Population Studies* 59, special supplement.

Scarman, Lord (1981), *The Brixton Disorders, 10–12 April 1981*, London: HMSO.

Seabrook, J. (1971), *City Close-Up*, London: Penguin.

Selbourne, D. (1983a), 'The new exodus', *New Society*, 19 May, 255–9.

Selbourne, D. (1983b), 'I'm getting out before it fall on me', *New Society*, 26 May, 295–7.

Shah, N.M. (1983), 'Pakistani workers in the Middle East: volume, trends, and consequences', *International Migration Review*, 17, 410–24.

Shah, S. (1979), 'Who are the Jains?' *New Community*, 7, 369–75.

Short, J.R. and Bassett, K. (1978), 'Housing Action Areas: an evaluation', *Area*, 10, 153–8.

Short, J.R. and Bassett, K. (1980), *Housing and Residential Structure: Alternative Approaches*, London: Routledge & Kegan Paul.

Simpson, A. (1981), *Stacking the Decks: A Study of Race, Inequality and Council Housing in Nottingham*, Nottingham: Nottingham Community Relations Council.

Siu, P.C.P (1952), 'The sojourner', *American Journal of Sociology*, 58, 34–44.

Sjoberg, G. (1960), *The Pre-industrial City: Past and Present*, New York: Free Press.

Skellington, R. (1981), 'How blacks lose out in council housing,' *New Society*, 29 January, 187–9.

Smith, D.J. (1974), *Racial Disadvantage in Employment*, London: Political and Economic Planning.

Smith, D.J. (1976), *The Facts of Racial Disadvantage*, London: Political and Economic Planning.

Smith, D.J. (1977), *Racial Disadvantage in Britain*, Harmondsworth, Penguin.

Smith, D.J. and McIntosh, N. (1974), *The Extent of Racial Discrimination*, London: Political and Economic Planning.

Smith, D.J. and Whalley, A. (1975), *Racial Minorities and Public Housing*, London: Political and Economic Planning.

Smith, D.M. (1969), *The North West*, Newton Abbot: David and Charles.

Smith, G. (1984), 'The Patels of Britain', *Sunday Times*, 26 February, 33–4.

Smith, S.J. (1981), 'Negative interaction: crime in the inner city', in *Social Interaction and Ethnic Segregation* (eds. P. Jackson and S.J. Smith), London: Academic.

Srinivas, M.N. (1967), *Social Change in Modern India*, Berkeley: University of California Press.

Stevens, L., Karn, V., Davidson, E., and Stanley, A. (1981), *Ethnic Minorities and Building Society Lending in Leeds*, Leeds: Leeds Community Relations Council.

Stonequist, E.V. (1935), 'The Problems of the marginal man', *American Journal of Sociology*, 41, 1–12.

Swinerton, E.N., Kuepper, W.G., and Lackey, G.L. (1975), *Ugandan Asians in Great Britain,* London: Croom Helm.

Taeuber, K.E. and Taeuber, A.F. (1964), 'The negro as an immigrant group', *American Journal of Sociology*, 69, 374–82.

Tambs-Lyche, H. (1980), *The London Patidars*, London: Routledge & Kegan Paul.

Taylor, J.H. (1976), *The Half-Way Generation*, Windsor: National Foundation for Educational Research.

Taylor, S. (1981), 'Riots: some explanations', *New Community*, 9, 167–72.

Taylor, S. (1982), *The National Front in English Politics,* London: Macmillan.

Thompson, J. (1982), 'Differential fertility among ethnic minorities', in *Demography of Immigrants and Minority Groups in the United kingdom* (ed. D.A. Coleman), London: Academic.

Thompson, M.A. (1970), 'A Study of Generational Differences in Immigrant Groups with Particular Reference to Sikhs', M. Phil. thesis, University of London.

Thompson, M.A. (1974), 'The second generation—Punjabi or English?' *New Community*, 3, 242-8.

Tinker, H. (1977), *The Banyan Tree: Overseas Emigrants from India, Pakistan and Bangladesh*, Oxford: Oxford University Press.

Trudgill, P. (1974), *Sociolinguistics: An Introduction*, Harmondsworth: Penguin.

Twaddle, M. (1975), *Expulsion of a Minority: Essays on Ugandan Asians,* London: Athlone Press.

Uganda Resettlement Board (1974), *Final Report,* London: HMSO.

Walker, M. (1977), *The National Front,* Glasgow: Fontana.

Ward, R. (1975), 'Residential Succession and Race Relations in Moss Side, Manchester', Ph. D. thesis, University of Manchester.

Webber, R. and Craig, J. (1978), *Socio-Economic Classification of Local Authority Areas*, London: HMSO.

Werbner, P. (1979), 'Avoiding the ghetto: Pakistani migrants and settlement shifts in Manchester', *New Community*, 7, 376-89.

Werbner, P. (1980), 'From rags to riches: Manchester Pakistanis in the textile trade', *New Community*, 8, 84-96.

Williamson, W. and Byrne, D.S. (1979), 'Educational disadvantage in an urban setting', in *Social Problems and the City*, (eds. D.T. Herbert and D.M. Smith), Oxford: Oxford University Press.

Wilson, A. (1978), *Finding a Voice*, London: Virago.

Winchester, S.W.C. (1974), 'Immigrant areas in Coventry in 1971', *New Community*, 4, 97-104.

Wirth, L. (1928), *The Ghetto*, Chicago: University of Chicago Press.

Woods, R.I. (1979), 'Ethnic segregation in Birmingham in the 1960s and 1970s', *Ethnic and Racial Studies*, 2, 455-77.

Wright, P. (1964), 'Go-betweens in industry', *Institute of Race Relations Newsletter*, 28-30.

Wright, P. (1968), *The Coloured Worker in British Industry*. London: Oxford University Press.

Yancey, W.L., Ericksen, E.P., and Juliani, R.N. (1976), 'Emergent ethnicity: a review and reformulation,' *American Sociological Review,* 41, 391-403.

Yinger, J.M. (1981), 'Towards a theory of assimilation and dissimilation', *Ethnic and Racial Studies* 4, 249-64.

Author Index

67-99

Subject Index